Holocaust and Nature

Holocaust and Nature

Edited and Introduced
by
Didier Pollefeyt

LIT

Dedicated to Nikolaas and Florian,
Helena and Barbara

Cover Picture
Bergen-Belsen – Station 5 – 1994 – 47 ½" by 75" drawing in charcoal and white Conté accompanied 47 ½" by hand-lettered poem. Drawing and poem/drawing are framed in hand-forged wrought iron frames. Artist: Arie A. Galles.

Sponsored by The Stephen S. Weinstein Holocaust Symposium

Bibliographic information published by the Deutsche Nationalbibliothek
The Deutsche Nationalbibliothek lists this publication in the Deutsche Nationalbibliografie; detailed bibliographic data are available in the Internet at http://dnb.d-nb.de.

ISBN 978-3-643-90313-6

A catalogue record for this book is available from the British Library

Klosbachstr. 107
CH-8032 Zürich
Tel. +41 (0) 44-251 75 05
Fax +41 (0) 44-251 75 06
E-Mail: zuerich@lit-verlag.ch
http://www.lit-verlag.ch

LIT VERLAG Dr. W. Hopf
Berlin 2013
Fresnostr. 2
D-48159 Münster
Tel. +49 (0) 2 51-620 320
Fax +49 (0) 2 51-23 19 72
E-Mail: lit@lit-verlag.de
http://www.lit-verlag.de

Distribution:
In Germany: LIT Verlag Fresnostr. 2, D-48159 Münster
Tel. +49 (0) 2 51-620 32 22, Fax +49 (0) 2 51-922 60 99, E-mail: vertrieb@lit-verlag.de
In Austria: Medienlogistik Pichler-ÖBZ, e-mail: mlo@medien-logistik.at
In Switzerland: B + M Buch- und Medienvertrieb, e-mail: order@buch-medien.ch
In the UK: Global Book Marketing, e-mail: mo@centralbooks.com
In North America: International Specialized Book Services, e-mail: orders@isbs.com

Contents

Preface

Didier Pollefeyt

In this volume Holocaust scholars reflect on the challenges of the Holocaust for our contemporary relation to nature. Currently there are not many studies available that deal with the relation between the Holocaust and ecological issues. In the first place the Holocaust was a human catastrophe, an evil committed against humanity, not against nature. In the first decade of the twenty first century, the ecological crisis has become increasingly central within the public arena. The destruction of our natural environment represents a threat for the whole human race. In connection with the ecological crisis the expression 'ecological Holocaust' is often used, stressing the seriousness of the present condition of planet earth and the way we deal with it. However, at the same time this expression risks reducing the historical Holocaust perpetrated by Nazism (1933-1945) to just being a metaphor for evil, suffering and death in general. This book tries to connect the specific, moral drama of the historical Holocaust of the twentieth century with contemporary ecological issues. The authors in this volume reveal the many, complex and challenging connections that can be found between the historical Holocaust, the way the Nazis understood nature, the way the victims of the Holocaust experienced nature and the way we approach nature today both on an individual and collective level.

Connecting the topics of Holocaust and nature often meets resistance both from the side of people working in ecology and from people working in Holocaust and genocide studies. For ecologists, it is not always clear why we should put so much energy into analysing a genocide that happened more than sixty years ago while today the human race as a whole is endangered by an even greater catastrophe. They are also often concerned that bringing ecology and Nazism into too close a relation – because of the ecological concerns and interests of the Nazi's – could harm the ecological movement today. From the side of Holocaust and genocide studies objections and resistance also arise through questions like: "Why care so much about animals and animal rights when human rights are violated and continue to be violated to such a large extent?" Still another danger is connected with making comparisons too easily between the suffering of the victims of the Holocaust and the treatment of animals in modern technological food industries, at the cost of the uniqueness of the Holocaust. These complexities, and many times ambivalences, in the links between Holocaust and nature also belong to the object of this volume.

This book will study the relation between Holocaust and nature from the perspective of the victims, the perpetrators and the bystanders. For the victims of the Holocaust, nature was often a mixed blessing, a source of both pain and hope. Next to the Nazi atrocities, nature was often a source of additional pain and suffering (cold, hunger, disease, etc.). Many victims' testimonies witness to the indifference of nature regarding their suffering, or to the radical rupture between the beauty of nature and the trauma of genocidal evil. But nature was also often a source of comfort and hope. It offered a structure to keep oneself in life (time schedule, day and night, the rhythm of the seasons, etc.) or it was the source of aesthetic and even religious experiences. The relation of the perpetrators to nature can certainly be seen as complex and yet from a completely different angle, it was rather a perverted relation. On the one hand, Nazism claimed a close relation with nature, even glorifying "Blood and Soil" (*Blut und Boden*). It celebrated the contact of the German people with the land and it saw an almost mystical bond between the German land and German blood. On the other hand, the physical reality was depreciated and exploited mercilessly through modern technology, Nazi eugenics, medical experiments, gender manipulation, the destruction of the landscape by the construction of extermination camps etc. But most of all, the integrity of the earth was spoiled with the blood and bodies of the innocent victims. Further, the perspective of the bystanders also has relevance for observation and study here. We know from Holocaust studies how decisive the role of the bystanders was in making the Holocaust possible. Today, we see before our very eyes how the ecological crisis is developing. We are all bystanders, and some will say that we are even "all perpetrators". Can we learn something from the study of the Holocaust bystanders for dealing with the present ecological crisis? Today, we as bystanders develop analogous reactions to the bystanders of sixty years ago: "is it not exaggerated?", "this can never be true", "what can we do as individuals against such large scale problems", etc. During the Holocaust, some bystanders became rescuers through small individual actions or through political engagement. Can we learn something from them and can the lessons of the Holocaust make us tomorrow's rescuers of the earth?

Today we can no longer be ignorant about the negative effects of our actions on the natural world. John Roth considers the remembrance of the Holocaust as a resource that can aid and help us significantly in dealing with nature. In the first chapter of this volume in response to the question "What have we learned from the Holocaust?", he reflects on the answers "not enough" and "maybe something". The importance of thinking further about the content of this *something* is emphasised, but Roth is also aware of the risk involved in producing a new hope that is naïve and inadequate.

It was only with the development of modern Western science and technology, in the late nineteenth century, that humanity fully realized that they could take control of nature and manipulate it. Peter J. Haas asserts in chapter two that it was various medical discoveries that actually lay behind the Nazi attempt to create a new natural, and so social, order. It was exactly this possibility to control nature that created the foundation for the Holocaust. More concretely, Haas shows how the Nazis' intermingling of racial theory, social Darwinism, eugenics and expanding medical knowledge contributed to the Holocaust. In fact, the incorporation of eugenic programs, and subsequent genocide, into Nazi public policy can be understood as the practical applications of medical science, as they understood it, to the social problems of post World War I Germany.

The embeddedness of – and ambiguity towards – nature in Nazi ideology is meticulously articulated in Margaret Brearley's essay in this volume (chapter three). The notion that Nature is the source of human power and creativity and must be preserved in its pristine state was counterbalanced, in Nazi principles, by the often radical exploitation of natural resources for military or even artistic purposes. Similarly, the Nazi vision of the perfect Aryan body trained to fight for – or biologically reproduce – that ideal was offset by the willing risk of mass death in submission to that very ideal.

Throughout history, the anti-Jewish legend of the wandering Jew was used to describe the Jewish situation in the Diaspora. In her chapter, Rochelle L. Millen describes the centrality of the land of Israel according to biblical and rabbinic sources and the interpretations of the medieval thinker Moses Nahmanides. She compares these with the Christian interpretations of the same passages, concluding with the vision of the Enlightenment on Jews and nationhood. According to Millen those Christian and rational outlooks were the immediate cause for both the Holocaust and for the famous Zionist theories on nature and the land of Israel by Moses Hess, Leon Pinsker and Ahad Ha'am.

Today's environmentalism isn't as innocent as it seems. In chapter five, David Patterson draws attention to some shocking likenesses it shares with the Nazi vision on nature: both are pagan points of view, solely interested in nature as an end in itself. Instead of the voice of Nature, Patterson offers an alternative one: the voice *beyond* Nature: the voice of G-d, seen through the eyes of Jewish mysticism. The environment has become the concern of the future, but Patterson reminds us to keep an eye on the past.

In the Bible humanity's relationship to nature is not just a matter of oppression, as is often believed. Although not only humanity, but the whole of creation has been taken up in the covenant with God, it is only humanity who has been called on to take responsibility for nature's survival. A hermeneutical openness to nature as God's creation is needed. The meaning of nature as a Trace of God's creation will only present itself when humanity is able to reserve a space within itself for the other as other. And according to the Jewish philosopher Catherine Chalier, it is this disinterestedness that should contain the key to a new, ethical relationship with nature. Even *after* Auschwitz, Didier Pollefeyt argues in chapter six, we are able to perceive nature as a work of God and to relate with it in a morally responsible way.

For Sarah K. Pinnock, the majority of Jewish authors research the Holocaust and nature from a male and patriarchal perspective. In her chapter she shows the other side of the coin through feminist theories and the testimonies of female Holocaust victims. Her focus on female authors is motivated by a desire to include women's voices in the dialogue on Holocaust and nature. She therefore deals with the work of two female Jewish thinkers whose writings relate directly to these themes: Simone Weil and Melissa Raphael. In the work of these authors she hopes to find a more holistic view on humanity and nature.

This book as a whole is constructed in the form of a dialogue among the authors. Every chapter is followed by responses from two other authors and then a subsequent response by the original author.

The last chapter describes Arie Galles' view on the relationship of Holocaust and nature. As a child, Galles grew up in post-Nazi Poland, living among the ruins left by the *Third Reich*. Many years later, in 1993, while visiting a local Holocaust Memorial Centre, the same experiences came back again; leading to his work of art called 'Fourteen Stages' or '*Hey Yud Daled*', fourteen aerial views of extermination camps, painted in charcoal. They give strong indications of how Nazism violated both humans and the landscape. As Arie Galles writes on his website, although art can't express the Holocaust, it would be to assign victory to the perpetrators if we were ever to withdraw art from confronting this horror.[1] The cover of this book presents Stage Five and is his artistic aerial presentation

---

[1] Arie A. Galles, *Fourteen Stages: Hey Yud Daled* on his website http://www.ariegalles.com/fourteen-stations.html (accessed March 2010).

of the camp of Bergen-Belsen. This book also ends with work from Galles, more specifically a fragment from his written diaries, where the artist describes his experiences, thoughts and feelings while making Stage Five of his masterpiece. The title of this dairy fragment is 'Skull' referring to the skull that appeared in Stage Five on the center right of the camp, just outside the fence (see the cover of this book), in front of the eyes of the artist while he was drawing. Galles' areal perspective on the camp of Bergen-Belsen shows in a moving and shocking way how the Nazis not only destroyed the moral landscape but also the natural landscape. The skull in the drawing and on the cover of this book shows in an artistic way how even nature reveals the deathly character of Nazism and protests against its genocidal politics. In his diary at the end of this volume, Galles writes: "Is Nature screaming to heavens the nature of this place?"

The World around Us:

What Have We Learned from the Holocaust?

John K. Roth

It is evident in the world around us that very dramatic changes are taking place.

Al Gore, *An Inconvenient Truth*

The statement from Al Gore that serves as the epigraph for this essay refers to the impact of global warming and climate change on the world around us. Long in the making, and largely owing to human decision-making, that impact portends eventual disaster for humankind. Not all of the devastation that human beings have wreaked on the natural environment that is our home can be identified as intentional, but increasingly it is an inconvenient truth that men and women can no longer claim to be ignorant about the negative effects of our actions on the natural world. In Gore's words: "very dramatic changes are taking place."[1] As far as human existence is concerned, the changes that are most prominently affecting the environment are not good. There are chances for human beings to intervene against that trend, perhaps even to reverse it, but the odds are not robustly in our favour.

A further inconvenient truth is not only that very dramatic changes are taking place in the twenty-first century, but also that such changes have been taking place for centuries. One of them remains the Holocaust, the systematic, state-organized persecution and murder of approximately six million Jews by Nazi Germany and its allies and collaborators. At first glance, links may not be obvious between Nazi Germany's genocidal actions against the Jews in the 1930s and 1940s and the environmental crisis that is unavoidably and increasingly at the center of human attention less than a century later. Second and third glances, however, are in order where these two realities are concerned. Here I will use a question – *what have we learned from the Holocaust?* – to explore some of them.

One thing that can be learned from the Holocaust is that many Jews – the murdered ones as well as those who survived – wrote about what happened to them

[1] Al Gore, *An Inconvenient Truth: The Planetary Emergency of Global Warming and What We Can Do about It* (New York: Rodale, 2006), p. 42.

and their families. As those diaries and memoirs describe an unfolding and unrelenting catastrophe, they frequently include observations about the seasons of the year, the weather, the sky, and other features of the natural world. *Night*, Elie Wiesel's famous and widely read Holocaust memoir, provides one powerful example. Narrating his experiences as the Holocaust engulfed Wiesel's family and community, that text makes repeated note of small but significant details of this kind as the seasonal cycle moved from the spring of 1944 to the spring of 1945. A listing of Wiesel's comments, almost poetic in their combination, could look as follows:

> Spring 1944. [...] The trees were in bloom.
> The eight days of Passover. [...] The weather was sublime.
> Some two weeks before Shavuot. A sunny spring day [...]
> All this under a magnificent blue sky.
> A summer sun.
> The lucky ones found themselves near a window; they could watch the blooming countryside flit by.
> The night had passed completely. The morning star shone in the sky.
> It was a beautiful day in May. The fragrances of spring were in the air. The sun was setting.
> "All of creation bears witness to the Greatness of God."
> Winter had arrived.
> An icy wind was blowing violently.
> Snow was falling heavily.[2]

Taken by themselves in this listing, removed as the individual descriptions are from the context in which they are embedded, these references to the natural environment are entirely ordinary. In fact, however, they create an extraordinary contrast and backdrop to the events that make *Night* as ominous as it is unforgettable. One more brief passage from Wiesel's testimony casts this reality in a particularly powerful way: "Never shall I forget," writes Wiesel, "the small faces of the children whose bodies I saw transformed into smoke under a silent sky."[3] That sentence is part of a tipping point in *Night*, for upon Wiesel's entry to Birkenau, the part of Auschwitz that was both the arrival point for his deportation transport and the main killing center in the vast Auschwitz camp complex, he was forever changed as he saw children, alive and dead, thrown into flaming pits and consumed. "Never shall I forget," he repeats seven times. "Never shall I

---

[2] Elie Wiesel, *Night*, trans. Marion Wiesel (New York, NJ: Hill and Wang, 2006), pp. 8, 10, 12, 15, 17, 23, 37, 40, 68, 77, 85, 96.
[3] Ibid., p. 34.

forget those things, even were I condemned to live as long as God Himself. Never."[4]

In one way or another, each of *Night*'s descriptions of nature anticipates, foreshadows, or helps to recall what Wiesel cannot forget and what human beings should remember as a warning. There are at least three dimensions to these relationships as they are found in Wiesel's memoir and in those of other Holocaust victims and survivors. First, the beauty of nature contrasts markedly with the brutality that human beings have inflicted, which included cattle-car deportations to human-created places of human-created degradation, filth, starvation, disease, death-dealing labor, violence, murder, and graves unmarked or nonexistent. Second, in *Night* and other Holocaust memoirs, nature sometimes appears to be conspiratorial in compounding the suffering that the Jews experienced at the hands of their German captors. Particularly the extreme heat of summer or the icy cold of winter diminished life chances for those who had no protection against the weather's extremes. Even more pronounced, a third implication in *Night* is that nature was indifferent to Jewish plight during the Holocaust. The sun rose and set, night came and went, the stars shone and the moon beamed, the seasons passed, the earth stayed in its orbit while devastation raged and seemed to make no difference in nature's order.

It is possible that a fourth dimension exists in these three relationships, for much of what needs to be learned from the Holocaust will be missed if we overlook how that event did and should affect the natural environment that is our home. To see more of how that fourth dimension might be mapped, this essay's response to the question 'What have we learned from the Holocaust?' moves on in two directions: (1) *maybe something*, but (2) *not enough*. To explain what I mean by those two phrases, and to identify their content, it is important to consider the governing question in some further detail.

There could be many questions with the form 'What have we learned from . . .?' The 'blank' could be filled by references to fields of inquiry, such as science or economics, or to events, such as some recent election or the 'war on terror'. The 'blank' could also be filled by references to persons, such as Moses, Jesus, or one might add contemporary political leaders such as George W. Bush or Osama bin Laden. The version of the topic question under consideration here, however, asks what we may have learned from the Holocaust, and at the very least it implies that there is something of particular importance about *that* event. One way

[4] Ibid.

to consider this importance is to observe that the time of this writing in the spring of 2007 was close to three significant Holocaust-related anniversary dates.

First, note that Irena Klepfisz, a child survivor of the Holocaust, became an important writer and poet whose best known works include a poem called *Bashert*. Its title, a Yiddish word, evokes senses of inevitability and fate. As the poem expresses grief and protest, *Bashert* does not mention the Holocaust directly, but that event shadows every line: "These words are dedicated to those who survived," writes Klepfisz, "these words are dedicated to those who died."[5]

Klepfisz's father, Michael, was one of those who died. Yisrael Gutman, survivor and historian of the Warsaw ghetto uprising, says that Michael Klepfisz, "who played an important role in the manufacture of armaments in the ghetto," was killed on the afternoon of April 20, 1943, in hand-to-hand fighting against General Jürgen Stroop's German forces, who had been directed to destroy the ghetto.[6] In the summer of 1942, the Germans had deported 300,000 Jews from the Warsaw ghetto. Most were sent to their deaths in Treblinka's gas chambers. Michael Klepfisz and Yisrael Gutman were members of a group of young Jewish men and women (they numbered about 750) who trained and armed themselves as best they could and were determined to resist Nazi efforts to annihilate the 55,000 Jews who remained in the ghetto in the early spring of 1943.

In January of that year, SS chief Heinrich Himmler had ordered further deportations, but Jewish resistance impeded that effort. Aware that they would meet determined resistance, the Germans regrouped and returned to finish the job on Monday, April 19, the eve of Passover in 1943. What ensued was, in Gutman's words, "the first urban uprising in German-occupied Europe, and, among the Jewish uprisings, the one that lasted the longest, from April 19 to May 16, 1943." The inadequately armed Jewish fighters, who also lacked military training and battle experience, were outnumbered three-to-one by Nazi forces that enjoyed support by tanks and cannons. The Jews' 'arsenal' consisted mainly of pistols, Molotov cocktails, and a few rifles, which had been smuggled into the

---

[5] For more information on Irena Klepfisz, see Carol Rittner and John K. Roth, eds., *Different Voices: Women and the Holocaust* (St. Paul, MN: Paragon House, 1993), pp. 324-27.

[6] For more detail on these matters, see Yisrael Gutman, *The Jews of Warsaw 1939-1943* (Bloomington, IN: Indiana University Press, 1989), pp. 370-400, especially pp. 378ff. See also Gutman's article on the Warsaw ghetto uprising in Israel Gutman, ed., *Encyclopedia of the Holocaust*, 4 vols. (New York, NJ: Macmillan, 1990), pp. 1625-32.

ghetto or taken from Germans who were ambushed in the previous January's resistance.

Using hit-and-run tactics, taking advantage of bunker hiding places, the Jewish fighters kept Stroop's forces off balance during the uprising's early days, but the Germans retaliated by burning the ghetto, building by building. Even then, doomed though it was, Jewish resistance continued. Not until May 8 did the Germans capture the Jewish Fighting Organization's headquarters, a bunker at 18 Mila Street. Mordecai Anielewicz, the Organization's commander, perished in that struggle. On May 16, Stroop declared victory, proclaiming that "the Jewish quarter of Warsaw no longer exists." The losses the Jewish fighters inflicted on the Germans were militarily small. Stroop's report noted 16 dead and 85 wounded. Nevertheless, the Warsaw ghetto uprising remains an immensely important example of heroic Jewish resistance against the Holocaust's overwhelming odds.

About four years later, on April 16, 1947, a man very different from Michael Klepfisz also lost his life. On that date, Rudolf Höss was executed by hanging on a gallows that stood near one of the gas chambers at Auschwitz, a site that has become nearly synonymous with the Holocaust itself. Höss, a Nazi SS officer, had been the commandant of that slave labor and killing center on Polish soil, which has rightly been called the epicenter of the Holocaust, for more than a million Jews were murdered there and tens of thousands of Polish prisoners also died in that place. Höss's execution came too late. However just it may have been, it could not bring back the dead. Nor could Höss's hanging repair the damage unleashed by his actions and the antisemitism and racism that provoked and inflamed them.

A third Holocaust-related anniversary worth remembering at the time of this writing is May 14, 1948. On that date the modern State of Israel came into existence. Survivors of the Holocaust were prominent among its first citizens, as they remain in their dwindling numbers sixty years on. The event we call the Holocaust was Nazi Germany's attempt to destroy Jewish life root and branch. Fortunately, that aim was not successful, at least not entirely, but it came far too close for comfort. However much the Holocaust was a factor in making the State of Israel a reality, no sound judgment is likely to claim that the Holocaust was a price worth paying for Israel's existence. Too much was lost to justify even that outcome.

As the three Holocaust-related anniversaries may help to make clear, the Holocaust was not only *genocide* but also a form of genocide that was *unprecedented*, owing to its systematic, unrelenting, technological, geographical, and ideological scope. For those reasons, it seemed possible that *we* (humankind) might learn something from this distinctive disaster, something that would have ethical, political, and even ecological importance.

Survivors of the Holocaust, at least some of them, thought that such learning might take place if people would *listen* to what had happened to them. In addition, as education about the Holocaust began to be more pronounced, a similar hope was echoed, namely, if people *studied* the Holocaust, the world might become, as we are fond of saying, 'a better place'. Unfortunately, those hopes were too optimistic. Ethnic cleansing in the former Yugoslavia, genocide in Rwanda, a humanitarian crisis of genocidal proportions in Darfur, Sudan, are just a few reminders that the slogan 'Never again' is less than credible, a fact that led the Holocaust scholar Hank Knight, a song-writing friend of mine, to compose a piece called *Hardly Ever Again*. Its apt lyrics go like this:

In '45, remember when
The world said, 'Never, never again!
Never again: six million lost;
Never again: the Holocaust.'
'Never,' we said, 'Never again.'
But this is now and that was then.
'Hardly ever again.'
Is that what we meant to say?
'Hardly ever again.'
Will we turn and walk away?
This is now and that was then;
And we meant 'hardly ever again.'
But this is now and that was then.
When will we ever mean 'never again'?

On a less overtly lethal but still immensely problematic note, *we* (humankind) have not even taken sufficiently to heart the harm that words can do. Absent slurs and slanders and the stereotyping that typically accompanies them, the Holocaust and genocide are scarcely imaginable, but the damage produced by ill-meaning words, spoken intentionally or just thoughtlessly or for some so-called 'shock effect' can be incalculable, a fact that the 2007 scandal surrounding the American talk-radio star, Don Imus, helped to make clear.[7]

Not enough, then, is a response, a sobering one, that must be made to the question 'What have we learned from the Holocaust?' Fortunately, if we not only recognize but also take to heart the fact that *not enough* has been learned, then it could be that the outcome is not entirely bleak. Knowing that we lack something, that we fall short, even that we might do better, is *maybe something* to consider as we keep asking 'What have we learned from the Holocaust?'

At this point, however, caution is advisable. Specifically, the turn from *not enough* to *maybe something* ought not to be so smooth and seamless that it produces a renewed hope that might turn out to be as naive and inadequate as the one that thought attention to the Holocaust would be sufficient to produce credible versions of the slogan 'Never again.' At the very least, we ought to think further about the content of the *something* that might be learned from attending to the Holocaust and its reverberations. With that aim in mind, at least two basic matters are among those that bear watching.

First, people need to take even more seriously than has been the case thus far that *what happened, happened*, specifically that the Holocaust really happened. The worst that can happen may seem unthinkable, but it is not impossible. Human beings are not able to prevent every catastrophe, but unless we are mindful about what can happen and how our actions have vital parts to play, we are less than responsible, and we put human life and its fragile environment needlessly in jeopardy.

Second, while our wordl may be a world where contemporary concerns (global warming, for instance, or the ongoing war against terrorists) loom so large that attention to the Holocaust sometimes seems beside the point, it remains the case that memory of the Holocaust is something that we ignore at our peril. If we are faithful to that memory, it enjoins us to bear witness for the living and the dead in ways that protest against and resist any and all of the forces that lay waste to human well-being. This protest and resistance will be less than fully responsible if it does not recognize and respond effectively to the fact that caring for our natural environment is crucial. Conversely, as three contemporary events help to

---

[7] As popular as he is controversial, Imus was fired by CBS in April 2007 after his racial slurs about the Rutgers University women's basketball team which provoked a firestorm of controversy. By the end of 2007, however, Imus was back on the air with another network.

show, if we fail to remember the Holocaust, we lose a resource that can aid and help us significantly.

In 2007, the observance of Earth Day took place on April 22. On that occasion, consciousness about global warming's perils and our accountability for causing them and, if possible, for overcoming them was upon us more than ever. The forecasts can be so grim and daunting as to make us think that attention to the past, even to the Holocaust, would involve a diversion of attention and energy from the world-saving responsibility that is before us. On the contrary, however, in such a case it might help us (at least some of us, some of the time) to remember Michael Klepfisz and all the others who resisted the great odds against them in the Warsaw ghetto. *They did not despair; instead they gave all they had to battle against human actions that were destructive of life and that produced a world ecologically as well as politically and spiritually scarred by its killing centers, mass graves, and incinerated bodies whose smoke and ash continue to cloud and shadow our environment.*

Disrespect for human life and disregard for the natural world are intertwined. Usually perpetrated under the cover of war, genocide, including the Holocaust, wreaks havoc on the natural world as well as on human life itself. Conversely, where human life is fully respected, the likelihood that the natural world will be better cared for is enhanced. Where the natural world is cared for well, it is also likely – although, unfortunately, not automatically guaranteed – that respect for human life will be high.[8] Michael Klepfisz and his sisters and brothers in resis-

---

[8] A Holocaust-related point requires the qualification in this sentence. I refer to the fact that, at least before the hell of World War II and the Holocaust broke out in, all of its fury, Nazi Germany included environmental concerns. Debates continue about 'how green were the Nazis?' Such as they were, the Nazis' conservation and ecological interests were neither unified nor unifying. Nazi plans for the protection of nature were more sporadic than sustained, and they reflected infighting more than coherence. Far from being global, Nazi environmentalism also tended to be "local, regional, or state-centered". See Franz-Josef Brüggemeier, Mark Cioc and Thomas Zeller, eds., *How Green Were the Nazis? Nature, Environment, and Nation in the Third Reich* (Athens, OH: Ohio University Press, 2005). The quoted phrase is from the editors' introduction to this volume, p. 2. Also helpful in this regard is Frank Uekoetter, *The Green and the Brown: A History of Conservation in Nazi Germany* (Cambridge: Cambridge University Press, 2006). I continue to believe that a full-fledged, global caring for the natural world would be utterly inconsistent with – indeed in determined opposition to – genocide and the destructive warfare that so often cloaks it. Some caution, however, remains appropriate. By no means is every form of environmental concern incompatible with massive human rights abuses and even genocide itself.

tance can help us to see those interconnections, if only we do not forget who they were and what they did.

In the spring of 2007, Americans were shocked by the mass killings that took place on Monday, April 16, at Virginia Polytechnic Institution and State University (Virginia Tech, as it is usually called) in Blacksburg, Virginia. Schools, colleges, and universities are not the safe places that we have hoped, if not assumed, they would be. One may ask what good could it do, is it even appropriate, to spend time remembering and reflecting on Auschwitz when the disaster at Blacksburg was so close at hand. Seung-Hui Cho, the Virginia Tech student killer, was no Rudolf Höss, but like Höss and all of us, that person was an individual embedded in a society. Just as Höss neither administered Auschwitz single-handedly nor accounted for the Holocaust by himself, the Virginia Tech killer, who, like Höss, does bear responsibility for many deaths, was also embedded in a society.

When we study the Holocaust, we pay attention to individual perpetrators, such as Höss, but we also have to explore the social context and its pathologies, which may help to show how there came to be a man like Höss. *The right kind of memory of the Holocaust, the kind that probes history deeply, can help us to see that responsible responses – not only to the loss and grief surrounding events such as those at Virginia Tech in the spring of 2007 but also to the portents underscored in the current environmental crisis – must attend to the social and economic order from which they emerge and to what might be done to make it less likely that such devastation will recur or continue.* As we move into uncertain futures, memory of the Holocaust has perspective and guidance to give us if we remember well and do not lapse into forgetfulness.

One of those murdered at Virginia Tech was Liviu Librescu, a Romanian-born Holocaust survivor, who had been a professor of engineering at Tel Aviv University in Israel until a 1985 sabbatical year took him to Virginia Tech, which became his adopted home. Professor Librescu lost his life while blocking the door to his classroom with his body so that his students could escape the onslaught. Only a few hours before the shooting started and Professor Librescu was killed, a two-minute siren wailed across the State of Israel. In 2007, that wailing might have anticipated Professor Librescu's death and that of his fellow victims at Virginia Tech. Be that as it may, since 1951, Israelis have observed Yom Hashoah, the annual Holocaust commemoration that brings the country to a standstill to remember the millions killed by Nazi Germany, its allies and collaborators. Significantly, the day falls midway between the time of the Warsaw ghetto uprising and Israeli Independence Day.

The Israeli siren's scream expresses both grief and warning, especially at a time when voices such as those of the Iranian president Mahmoud Ahmadinejad deny the Holocaust's reality and even threaten to remove Israel from the face of the earth. Just as the Holocaust siren in Israel is a reminder of what was lost during the Nazi years and their reverberations, it warns that Israel's future is not guaranteed either but depends on vigilance and integrity. *It seems to me, moreover, that the particularity of Israel's siren on Yom Hashoah has a significant universality as well, one that warns against taking anything good for granted, including a life such as Professor Librescu's and the natural world that is our human home, a point that we all do well to take to heart and that memory of the Holocaust keeps driving home again and again in distinctive ways.*

What happened, happened. Do not despair, even against steep odds. Keep attention focused not only on individuals but also on contexts as we try to create a world less prone to violence and environmental decay. Take nothing good for granted. These four insights are important ones that could be part of the *maybe something* that keeps the *not enough* from being dominant when possible responses to the question 'What have we learned from the Holocaust?' are concerned.

To those four points, I would add one more as a kind of postscript: It is important to recognize that even small deeds and modest actions can be life-saving and environment-protecting. Particularly with that theme in mind, it can be appropriate to close these reflections about *not enough* and *maybe something* in response to the question 'What have we learned from the Holocaust?' by referring to the late American Jewish philosopher Philip Hallie, who was a distinguished professor at Wesleyan University in Connecticut.

As an artillery officer in World War II, Hallie helped to destroy Nazi Germany, but he was best known as an ethicist whose belief in the preciousness of human life led him to write about cruelty and also about a French village called Le Chambon, where five thousand Jews found a wartime haven in the 1940s. A fine storyteller, Hallie often recalled a hurricane that battered New England. The storm's devastating power gave him an apt metaphor for the human predicament. "It's the hurricane we're in," Hallie liked to say, amplifying his conviction with the admonition, "don't forget it."

Philip Hallie's concern about 'the hurricane' was not limited to lethal weather. Human existence, he thought, is always contending with hurricane seasons of one kind or another. Significantly, however, when Hallie tracked the hurricane that reached his Connecticut home, havoc was not all that he saw. As the storm raged, he noticed that there seemed to be space for calm and quiet within the hurricane's eye. Hallie's eye, moreover, was drawn to the pale blue sky overhead. That blue was his favorite color, for, as Hallie put it, his passion, his hope, was to "expand the blue." Some persons, he added, "make a larger space for blue, for peace, for love." Such work, he insisted, "takes power as well as love. It takes force of will. It takes assertion and commitment."[9] As we think about and act on the inconvenient truths embedded in the very dramatic environmental changes and challenges that are ours, as we think about and act on what has – and has not – been learned from the Holocaust, moving beyond *not enough* and adding what is needed to *maybe something*, depends on expanding the blue and on cultivating the qualities that are necessary for steps in those directions.

Closing on a personal note, I well remember one summer afternoon that I spent at Auschwitz-Birkenau in 1996. Late in the day, a thunderstorm drenched the remains of that camp. When the storm passed, the sun came out, and a magnificent rainbow arched over Auschwitz. The juxtaposition of Birkenau, the reality and aftereffects of that killing center, and a rainbow, with its awesome beauty and symbolic, even biblical, meanings of hope and promise, remains both jarring and poignant to me[10]. Memory of that intersection, even collision, between history and nature makes me wonder to what extent, indeed whether, humankind can significantly repair the damage we have inflicted on each other and on our natural world. That same memory enjoins me to say that failure to do our best in those regards is unacceptable.

---

[9] See Philip Paul Hallie, "Cruelty: The Empirical Evil," in Paul Woodruff and Harry A. Wilmer, eds., *Facing Evil: Light at the Core of Darkness* (La Salle, IL: Open Court, 1994), pp. 128-30. Also relevant is Philip Hallie, *In the Eye of the Hurricane: Tales of Good and Evil, Help and Harm* (Middletown, CT: Wesleyan University Press, 2001).

[10] As the relevant passage in the Hebrew Bible (Genesis 9:1-17) is usually interpreted, God uses the rainbow as a covenantal sign between "me and you and every living creature that is with you, for all future generations. I have set my bow in the clouds, and it shall be a sign of the covenant between me and the earth" (Genesis 9:13). Creation had nearly been destroyed entirely by what Genesis (6:5) calls "the wickedness of humankind" and the flood unleashed by God against that corruption. Noah, however, had found favor with God, and when the waters subsided, God promised him that "the waters shall never again become a flood to destroy all flesh [...] nor will I ever again destroy every living creature as I have done" (Genesis 9:15 and 8:21). If God's promises to Noah are problematic, and they are, it is also true that what Scripture calls "the wickedness of humankind" has caused and continues to produce incalculable harm and devastation in our world.

In Response to John K. Roth

Didier Pollefeyt

The chapter of John Roth reveals the many and complex relations between the study of the Holocaust and the contemporary ecological issues. From the perspective of the victims of the Holocaust, he explains how the beauty of nature was in a permanent contrast with the suffering and evil of Auschwitz. Not only was nature completely indifferent to the sufferings in Auschwitz, but, moreover, nature was sometimes a supplementary source of suffering and evil. In this presentation, nature had no hope to offer to the victims of the Holocaust. But if we read testimonies, we often see how nature was not only experienced as conspiratorial with the suffering in Auschwitz, but also sometimes as a source of hope. The beauty of nature remembered the victims of another reality in which beauty, freedom, colorfulness, life, vulnerability, etc. were revealed. Nature was a source of comfort in a world were everything was grey. The rhythms of nature gave orientation to the deadly uniformity of daily live in the camps in which there were no seasons, no Sunday or week day, no day and night, etc. Nature offered a calendar to the victims and in this way nature offered the experience of time, so crucial to be and to remain a human person. The dynamics of nature, finally, was also sometimes experienced as 'natural religion' in a context in which every formal, particular religion, especially Judaism, was forbidden. The dynamics of nature made clear the inherent and indestructible power of nature, and revealed the idea that after winter comes spring, always, that life is stronger then death. Often, believers lived their religious life (prayer, rituals, events, etc.) following the rhythms of nature. In the chapter of Roth, nature had little hope to offer to the victims of the Holocaust, but on the other hand, Roth is convinced that the Holocaust has hope to offer for the future of nature. He argues that because the Holocaust was unprecedented – given its systematic, technological, geographical and ideological scope – we might learn something from it that would have ecological importance. For Roth, there is no opposition between remembering the Holocaust and confronting new actual challenges. In this way, he gives a clear answer to those who would argue that we should not continue to put our energy in the study and the remembrance of the Holocaust because of the new ecological dangers we are facing today. He makes this clear in several points. "The worst that can happen may seem unthinkable, but it is not impossible." Indeed, the Holocaust was beyond (in)human imagination, but the drama *did* happen. The impossible is possible. In confrontation with the dramatic perspectives on the future of the ecological system called 'earth', we also often think that the impossible is not possible. We often relativise the ecological challenges, comforting ourselves with the idea that the situation is not as bad as we think, that we do not see or experience ourselves the decline of nature, that things are exaggerated, etc. Nature is as vulnerable for evil and destruction as our interhuman relations. Roth makes us clear that our moral approach of the Holocaust can be fruitful for our moral approach of the ecological crisis: both need to take the individual (micro level) and social (marcro level) elements into account. The moral message of the Holocaust is not one of historical fatalism. During the Holocaust, it was possible to resist evil, both from the side of the perpetrators and the victims. Also in confrontation with the ecological disasters, we are not just delivered to macro-economical powers, we have the possibility to resist and to change reality. But just as studying the Holocaust is not only studying the individual victims or perpetrators, so does the ecological crisis ask for attention to the whole social and political context. And the attention to the social and political dimension of evil does not undermine the responsibility of the individual. Even the smallest individual act to protect human life and/or nature can make a difference. In this way, Roth made the confrontation with the Holocaust fruitful for our ecological responsibilities today. The rupture between history and nature presented in the image of the rainbow over Auschwitz so becomes a source of human responsibility. The approach of Roth is strongly influenced by the presupposition that disrespect for human life and disregard for the natural world are intertwined. Even if he knows that respect for the natural world does not automatically guarantee respect for human life, the basic premise remains the continuity between respect for the human and the natural world. My central question: does Roth not presuppose this continuity too easily? Can't we also learn moral lessens from the Holocaust if we stress the continuity instead of relativising and even denying the connection between the respect of Nazi for the natural world and their respect for Jewish and, more in general, human life. My point is that the recognition of this continuity can also be a source of moral lessons from the Holocaust that are relevant for the ecological issues today. Can we learn from the Nazi history that respect for nature and disrespect for human life can go hand in hand? The confrontation with the Holocaust asks us to qualify our speaking about 'nature'. What do we mean when we speak about 'nature' and how do we appreciate nature. In this context, it is interesting to see how Richard Rubenstein and Emmanuel Levinas developed a common understanding *after Auschwitz* of nature. Rubenstein speaks of Nature as the cannibalistic Mother Earth and Levinas speaks about nature as the *il y a* ('there is'), the all-absorbing and chaotic experiences of an anonymous, impersonal power. For Rubenstein, 'Mother Earth' is written with capitals[1], because it is God or the Divine that manifests himself or itself through the material cosmos. This God of Nature is neither a calming nor a loving presence. Nazism teaches us that in confrontation with the powers of this Nature, we have finally no argument, no defense, no rights. In fact, Levinas' idea of nature as il y a ('there is') has almost the same meaning as Rubensteins understanding of Mother Earth. The difference however is that Rubenstein identifies God with Nature, while Levinas contrasts God with nature. Nature has no divine legitimacy or power. God is an 'otherwise than being': He reveals himself as a radical call for responsibility, especially when human beings are dehumanized by natural powers or by people whose ideologies that pretend to act legitimized by natural powers. One of the crucial lessons I have learned from the Holocaust, is that only in such a critical reading of nature as an anonymous, un-divine power, it will be possible to resist those regimes that glorify nature and have disrespect for fellow human beings. So, where Roth shows how the rupture between the beauty of nature and the evil of the Nazi's can be source of moral lessons after the Holocaust, I argue how the continuity between the Nazi glorification of the beauty of nature and the evil of the Nazi's could also be a source of moral ecological consciousness after Auschwitz.

---

[1] Richard Rubenstein, *Morality and Eros* (New York, NJ: McGraw-Hill, 1970), pp. 32-41.

In Response to John K. Roth

David Patterson

With his usual eloquence and insight, John Roth has sounded a warning to all of us. As ever, he has a sharp eye for significant details of the Holocaust and for their far-reaching ramifications. He has done an excellent job of connecting those details to the landscape of the contemporary world, including the natural, physical world we live in. He notes, for example, that "disrespect for human life and disregard for the natural world are intertwined." While this is a true statement, I would like to see Roth elaborate a bit more on *why* it is true. When those of us who are made of the earth despoil the earth, we desecrate our own souls, as the Nazis did when they cast the ashes of the Jewish people over the face of the earth. In their assault on the human face, they transformed the face of earth itself. I understand very well why Roth states: "Where the natural world is cared for well, it is also likely that respect for human life will be high." Here too, however, he could do more in the way of explaining why this is the case. It seems to me that it is the case only where there is some sense of a higher relation to a Creator who has sanctified creation, has fashioned human beings after the image and likeness of His own holiness, and has commanded us to bring more human souls into this world. Among the environmentalists who respond to the global warming crisis, there are some who blame not only human actions but also *human existence* for the damage done to creation. It is as if our very presence in the world were a stain upon nature (a position reminiscent of the Nazis' view of the Jews). Thus on 21 November 2007 the British newspaper *The Daily News* ran a story about a woman who had herself sterilized in order to avoid polluting the planet with more carbon footprints. This offering up of human life for the sake of nature smacks of a paganism that plays into the hands of those who would engage in mass murder. For here an apparent *respect* for the natural world is intertwined with a contempt for humankind.

Asking what we have learned from the Holocaust, then, we must also ask what we have learned from global warming and other forms of polluting the planet. Have we learned that we must sterilize ourselves, that human existence is inherently evil, and that the earth would be a better place without all these carbon footprints defacing it? If that is the logical conclusion, then "the beauty of nature contrasts markedly" not just "with the brutality that human beings have inflicted," as Roth says, but also with the very existence of human beings, from Heinrich Himmler to Mother Teresa. When environmentalist movements are severed from the higher relation that sanctifies the human relation, they may resemble the Nazis more than they would like to imagine.

Consider, for instance, the following statement: "We recognize that separating humanity from nature leads to humankind's own destruction. Only through a reintegration of humanity into the whole of nature can our people be made stronger. That is the fundamental point of the biological tasks of our age. Humankind alone is no longer the focus of thought, but rather life as a whole." These words were written not by a representative of the Green Party, but by Nazi botanist Ernst Lehmann, who completes his thought by saying: "This striving toward connectedness with the totality of life, with nature itself, a nature into which we are born is the deepest meaning and the true essence of National Socialist thought."[1] The danger is that at times it may also become the true essence of postmodern environmentalist thought.

In the time of the Holocaust, as Roth points out, "nature sometimes appears to be conspiratorial in compounding the suffering that the Jews experienced at the hands of their German captors." And: "The sun rose and set, night came and went, the stars shone and the moon beamed, the seasons passed, the earth stayed in its orbit while devastation raged and seemed to make no difference in nature's order." And yet in their creation of the anti-world – of an anti-nature – the Nazis transformed nature into a wilderness of stark indifference that mirrored the indifference of humanity toward the extermination of the Jews. In case after case, victims of the Holocaust recount how that indifference added to their suffering. To be sure, in many Holocaust memoirs survivors describe a 'natural' surrounding that has no grass, no birds, no sun, no sky. In the anti-world nature itself was transformed.

Roth ends his essay by turning his eyes – and ours – toward the sky, toward the dimension of height that came under assault in the Shoah. And we see a rainbow. In Judaism we are taught to remember that dimension of height, of holiness. For upon seeing a rainbow, we bless God as the one "who remembers His covenant, is true to His covenant, and keeps it according to His word" (*...zokher habrit, veneeman bivrito, vekayam bemaaro*). And yet, when the rainbow appears over Auschwitz, Judaism regards the rainbow as a reminder of a great catastrophe, so

---

[1] Ernst Lehmann, *Biologischer Wille: Wege und Ziele biologischer Arbeit im neuen Reich* (Munich: Zentralverlag der NSDAP, 1934), pp. 10-11.

that we do not *gaze* upon it. It is the sign of a covenant between God and humanity, and because we live in a *covenant*, it is a sign of our obligation to cry out to God, even against God, in the name of the holiness of human life – a move that we cannot make as long as we think of human beings in terms of carbon footprints. For the rainbow is the sign of the promise of life and of the divine, *as well as the human*, sanctification of the life that comes forth from the earth.

Therefore in Judaism a rainbow symbolizes the holiness that links our treatment of one another to our treatment of creation. In the Book of Ezekiel, the prophet describes his ascent to the heavenly realms, where he glimpsed the hidden mysteries of the Divine. He concludes his description by saying: "Like the appearance of a bow which would be in the clouds on a rainy day, so was the appearance of the brilliance all around; it was the appearance of the likeness of the glory of God" (Ezekiel 1:28). The rainbow, then, symbolizes the presence of the Holy One Himself – even in the skies over Auschwitz, skies once transformed into a cemetery.

So we have this tension in the rainbow as a sign both of catastrophe and of divinity. It the time leading up to the flood, all of humanity – creation itself – had degenerated into a state of defilement. This desecration of human beings and the earth they inhabited, led the Creator to regret having created both. So He undertook a massive destruction not only of the human beings whose evil had corrupted creation, but also of the earth itself. One can see some parallels to the Holocaust. In the time of the flood the earth and humanity were destroyed by water, and in the time of the Holocaust they were purged by fire. However, whereas the flood had purified the earth and humanity, one wonders whether the catastrophe of the Shoah has purified anything, as Roth's question suggests: Have we learned anything?

When God set His rainbow in the heavens, it was a sign of his covenant with humanity and all of creation. And covenant is made of commandment: if He set his rainbow in the skies over Birkenau, it is a sign of what Emil Fackenheim calls "the Commanding Voice of Auschwitz." To heed this Commanding Voice is to live in the very covenant that the Nazis set out to obliterate. Therefore, if we are to refuse the Nazis a victory, despite their environmental consciousness, then we must attend to the care of our world and of the people who live in it according to the summons of the Commanding Voice, and not according to environmental sensitivity, love of nature, an appreciation of beauty, or any of the other reasons often invoked in environmentalist movements. It is not the song of the whales or the cry of the wolves that summons us – it is the Creator of the whales and the wolves, the One whose Voice was silenced at Auschwitz, when the clouds overhead were made of the ashes of the dead. For rainbows do not form in clouds of ashes.

Only where our care for the environment is undertaken as a response to its Creator, a regard for the natural world can be intertwined with a respect for human life. I believe this point is implied in Roth's essay. But it is a point that must be more than implied, lest we forget it ourselves and slip into a paganism that is perfectly compatible with purifying the land of its carbon footprints.

In Reply to My Respondents

John K. Roth

> Our chestnut tree is in full blossom. It is covered with leaves and is even more beautiful than last year.
>
> Anne Frank, May 13, 1944

Didier Pollefeyt and David Patterson are insightful interpreters and perceptive critics of *The World around Us*. I take their most telling questions to be the following: do I overemphasize the negative parts that the natural world played during the Holocaust and underestimate how the natural world provided hope to those targeted by Nazi genocide? Do I assume too much – and too easily – continuity between respect for human life and respect for the natural world? What circumstances are necessary to make claims about that continuity credible? I hope that the attention to details in my indirect route for addressing those questions will provide helpful responses.

An internet site linked to the solar panels on my house estimates that the average tree removes about seventeen pounds of carbon dioxide from the atmosphere each year. Even the modest energy output from my solar system can replace or add the equivalent of some five hundred trees annually. If humanity's carbon footprints are to be reduced, if global warming is to be checked and the environment on which human life depends is to be preserved, trees – forests of them – are of immense importance. Trees - forests of them – were of immense importance in Nazi Germany and eventually in the Holocaust too.

According to the geographer and environmentalist Michael Imort, the Nazi leader Hermann Göring, one of Adolf Hitler's most trusted lieutenants, became "the self-anointed *Reichsforstmeister* or 'Reich master of forestry,'" enforcing the *Dauerwald* policy, a term best translated, says Imort, as 'perpetual forest' or 'eternal forest.' Under this plan, ecologically advanced at the time, German foresters abandoned practices that "planted, thinned, and cut" simply to maximize economic yield. *Dauerwald* strategies focused not on the individual tree as a product but on the forest as an organism or ecosystem. Economic interests remained paramount and, in fact, were enhanced, but the stakes of the *Dauerwald* concept were higher than that because it emphasized not only sustainability but also cultivation of "*bodenständige* (native) species" and "exclusion of introduced 'foreign' species."

This forestry philosophy enabled Göring to proclaim: "Forest and people are much alike in the doctrines of National Socialism. (…) *Eternal forest and eternal nation are ideas that are indissolubly linked*" (Göring's emphasis). On December 13, 1934, the *Third Reich* anticipated those remarks, which Göring made about a year later, by establishing its Law Concerning the Protection of Racial Purity of Forest Plants. "Ask the trees," an influential forester would write in 1939, "they will teach you how to become National Socialists!"[1] In unintended ways, those words proved prophetic. The world war that Nazi Germany started that same year meant that vast forests in Nazi-occupied Poland, France and Norway, along with those in the expanded *Third Reich* itself, were exploited for military and genocidal purposes. Those purposes included production of railroad ties that facilitated transports to Auschwitz and lumber from which camp's primitive barracks were constructed to house Jews and others who were spared from the gas chambers but starved, were beaten, or worked to death.

Meanwhile, especially in Eastern Europe, the forest became a place where some Jews escaped, hid, and resisted as partisans. For them, trees meant hope during the Holocaust and even life itself. Thus, I welcome Pollefeyt's observation that it is too narrow to underscore only nature's seeming conspiracy with human-made atrocity, a point driven home all the more by reference to what is arguably the most important tree related to the Holocaust.

More than 150 years old, a horse chestnut tree still stands – at least at the time of this writing – not in a forest but in the inner garden of the house at Keizergracht 188 in the Dutch city of Amsterdam. Anne Frank could see that tree from the hiding place at Prinsengracht 263, where she and her family took refuge from July 6, 1942, until their betrayal on August 4, 1944, which resulted in Anne's deportation to Auschwitz and her death from typhus at Bergen-Belsen in early March 1945.

Several entries in Anne Frank's celebrated diary, including three poignant examples from 1944, focus on the Amsterdam chestnut tree and have made it famous, too. February 23: "The two of us looked out at the blue sky, the bare chestnut tree glistening with dew, the seagulls and other birds glinting with silver as they

---

[1] Michael Imort, "'Eternal Forest – Eternal *Volk*': The Rhetoric and Reality of National Socialist Forest Policy," in Franz-Josef Brüggemeier, Mark Cloc, and Thomas Zeller, eds., *How Green were the Nazis? Nature, Environment, and Nation in the Third Reich* (Athens, OH: Ohio University Press, 2005), pp. 43, 45, 52 - 54.

swooped through the air, and we were so moved and entranced that we couldn't speak." April 18: "April is glorious, not too hot and not too cold, with occasional light showers. Our chestnut tree is in leaf, and here and there you can already see a few small blossoms". May 13: "Our chestnut tree is in full blossom. It is covered with leaves and is even more beautiful than last year."[2]

In the early twenty-first century, schools around the world bear Anne Frank's name. Typically, they commit themselves to be centers of learning that highlight freedom, justice, human dignity, and Holocaust remembrance. Anne was born on June 12, 1929. In recent years, children from those schools have celebrated her birthday by planting seedlings germinated from the Amsterdam tree's chestnuts. That project took off when it became clear that the old tree, weakened by leakage of fuel oil from an underground storage tank and imperiled by fungicidal disease, had become a safety hazard. The tree could even endanger the secret annex where the Frank family hid. By the autumn of 2007, plans to axe the tree were far along. That November, however, Dutch officials and conservationists got a court injunction that spared the tree, which has since received special care to prolong its life for a decade or more. When the tree can't stand no more, and the time for its removal arrives, the planting of its clone may be possible.

Coupled with details about Nazi forestry and the parts that the woods could play in Jewish resistance and survival, the story of the Anne Frank Tree forms a kind of parable-response to the queries that Pollefeyt and Patterson raise. Just as it is true that the natural world can kindle and support hope where atrocity rages, even while nature in other guises may seem indifferent to or complicit in injustice and suffering, there is no easy continuity between respect for human life and respect for the natural world. The credibility of claims about the continuity between those sorts of respect depends on whether the right kinds of respect are in play. Everything hinges on their qualities.

The Nazis could be good foresters, but only up to a point. Their philosophy of forestry itself was intertwined with an ideology that was fundamentally disrespectful of human equality and human rights. They could not fully and truly be good foresters because their worldview – steeped in racism, militarism, and genocidal outcomes that had much to do with God-denying arrogance – entailed

[2] The quotations are taken from the Anne Frank House web site at: http://www.annefrank.org/content.asp?pid=445&lid=2. This site contains much information about the post-Holocaust saga involving the Anne Frank Tree, as it is now called.

that they would ravage forests, and much more, instead of saving, respecting, and nurturing them in ways that benefited humankind.

The genocidal catastrophe the Nazis unleashed, did immense harm to the natural world as it also took an incalculable toll on human existence and Jewish life in particular, including damage to the ways in which the natural world can appear to us. As Patterson suggests, a rainbow ought not to produce senses of irony and skepticism, unavoidable though such senses can be when one arches an Auschwitz sky. Instead, it ought to be a beautiful sign of hope and even covenant. It can be those things, if, among other things, we become fully and truly good foresters – figuratively and literally. One can glimpse what that means by paying further attention to Anne Frank's chestnut tree.

A tree is just a tree, of course, but that tree is not just any tree, and it may even help us to realize that no tree, indeed nothing in creation, is insignificant. A tree in Amsterdam did not and could not save Anne Frank's life, but it helped to sustain and inspire her. The Holocaust's perpetrators and the natural world's typhus killed her – the natural world is by no means always a friendly place – but those facts do not render meaningless, at least not entirely, the ethical sensibilities and spiritual convictions that her diary underscores.

One can be cynical about the passion that has been aroused to save the Anne Frank Tree. Those efforts do not ensure what is needed even more, namely concerted efforts to reduce needless suffering, to curtail human rights abuses, and to intervene against genocide, if not to prevent it. Nevertheless, in the determination to save a particular chestnut tree, one can identify elements of what is needed if humankind is to advance much larger and more desirable goals. Those elements include devotion to save life, dedication to resist the human and natural forces that destroy it, and determination not to give up when success seems to be a forlorn hope.

The continuity between respect for the natural and the human world is fragile, complex, and difficult. One rightly cares for and about the Anne Frank Tree and its seedlings both as a means for human good and as ends in themselves. Tension-filled though that relationship may sometimes be – especially when we think, mistakenly, that the natural order exists for human use and consumption alone –, the integrity of the natural world and the dignity of humankind, individually and communally, go hand in hand. They do so because human life cannot flourish without a healthy natural environment, and because a natural world in which human life is absent, abused or abusive with regard to itself and/or its environment, compromises and even lacks elements of awareness, responsibility, and awe on which the presence of sense and meaning within and about the natural world depends.

We cannot fully and truly respect human life without caring for the natural world, and we cannot fully and truly respect the natural world without caring for humanity both in individual and communal terms. We are part of the natural world, and the natural world is not fully and truly itself without us, partly because the natural world is itself in process. What we do within and for the world and ourselves makes a great deal of difference with regard to the order and meaning of existence itself. And, as Patterson rightly reminds us, if we listen well to the best that these two inseparable realms – the human and the natural – can say to us, neither of them exhausts reality, let alone completes it. They point toward, they even embody in ways that transcend themselves, at least traces of the divine. If we will let them, such traces can compel and focus properly our stewardship on earth.

The Medicalization of Nature

Manipulating Social Ecology

Peter J. Haas

The word 'nature' is derived from the Latin '*nasci*' (to be born) and refers to that which is inborn or innate. For most of human history, nature was in fact simply the given out there, with which people have had to contend. It is also the case that for the most part, the human encounter with this natural given has not been pleasant. In fact, nature has most often been described as hostile, capricious and even malevolent. It was only with the development of modern Western science and technology, that people for the first time felt that they could take control of nature and manipulate it to their own ends. In this regard, one of the most enticing areas of science was medicine. The medical sciences promised not only health, but even more electrifying, the possibility of controlling and maybe even overcoming death itself. Medicine in short, held out the hope that in some sense humans could replace the divine. My contention below is that it was precisely the medical discoveries of the late nineteenth century, in particular those dealing with the cause and spread of human pathologies, that lay behind the Nazi attempt to create a new natural, and so social, order. In particular, Nazi scientists and public policy administrators wanted to claim that they could shape the natural world in such a way as to create (or re-create?) the conditions under which the Aryan race would flourish. Further, medicine, with its promise to provide tools for engineering the human, gave Nazi policy makers a sense that they could transcend not only traditional Western religion but its accompanying morality, a morality which they felt had held humans in passive captivity for too long.

On a theological level, it is not hard to see the human triumph over nature as an apocalyptic, even messianic fulfillment of Western religion. According to Psalms, "The earth is the LORD's and all that it holds, the world and its inhabitants."[1] But we learn also that human beings almost immediately after their creation were given rulership over this realm. The first command recorded in the Bible was in fact: "Be fertile and increase, fill the earth *and master it; and rule the fish of the sea, the birds of the sky, and all the living things that creep on the*

[1] *Tanakh: The Holy Scriptures* (Philadelphia, PA/ New York, NJ: The Jewish Publication Society, 1988), Psalms 24:1.

earth."[2] The problem was that from the Hellenistic period forward, the material world of Creation was taken to be an imperfect realm full of decay, disease and death. For many thinkers in the Church the natural realm was so irretrievably evil that the only hope for the human soul was to be reborn into a new, better and eternal life in the spiritual realm. But it was also true that for many Christians and Jews, part of their religious heritage was about improving, and even redeeming this physical world through faithful obedience and so instituting in it the messianic Kingdom of God. In light of this theology, it is easy to see how the scientific advances of the Modern period could be taken as the ultimate divine blessing. The use of modern medicine in particular to overcome the natural curses of disease and death, would be nothing less than carrying out the ultimate salvific purposes of the divine.

It can not be overemphasized how much this Enlightenment optimism stood in contrast to the prior historical experience of humankind. Life as experienced by the vast majority of people had been, and still is, full of conflict, struggle, pain, poverty, suffering and death. Explaining why this is so, has been an issue for all religions. For the Western monotheistic religions, attempts to explain the evilness of the divine creation (theodicy) have often been reduced to describing nature as the arena in which divine punishments are meted out on a sinning humanity. According to this view, floods, droughts, plagues, disease and death are part of the natural order only insofar as they reflect divine responses to human imperfection and perfidy. Human control was only possible, if imperfectly, through rituals of confession, repentance and obedience. Consider, for example, responses in Europe to the Black Death that swept across Europe in the mid-fourteenth century. In the face of the utter horror of the pandemic, it was simply taken for granted that people were witnessing a punishment from God which could be stopped only by rituals of repentance and penance. The following account is one of many that illustrates this attitude.

> Believing that the Black Death was God's punishment upon man for the sins of the world, many people stopped swearing and gambling. Dice were changed into rosary beads. Huge religious processions were held. All over Europe, groups of men called flagellants, passed from town to town imitating Christ's martyrdom in the hopes of obtaining God's forgiveness. They would march for 33 and a half days because that marked one day for each year Christ was on

[2] Ibid., Genesis 1: 28; emphasis added.

> earth. While marching, they would scourge themselves and each other with leather whips tipped with iron spikes. Spectators would cry and howl and tear their own hair out. People would try to catch the flagellants' blood on their handkerchiefs, believing that it had the power to work miracles.[3]

In contrast to this traditional attitude, the Enlightenment proposed that humankind was entering an age in which the application of science and technology would make possible the rational management of nature. To be sure, there was still a strong undercurrent in the West that held that (to cite the title of a painting by Sir Edwin Landseer), "man proposes, nature disposes." Nineteenth century Romanticism in particular stressed the unknowable and mysterious forces of nature. But powerful scientific interests were able nonetheless to convince the broad public that human technical advances were going to be able more and more to overcome whatever obstacles nature seemed to place in the path toward human happiness and physical salvation. By the late nineteenth century, projects from vaccinations to railroad bridges to Chicago skyscrapers bore testimony to the human ability to triumph over the limitations once imposed by nature. Nature was no longer the arena of the divine in which humans passively absorbed their punishments, but an arena for human initiative, action and progress. It was thus fully in the spirit of the day that the Nazis claimed that they could take control over nature and manipulate it with the tools provided by science so as to serve their own national and political ends. In fact, they argued, it was their rational right, even moral duty, to do so. The result was, as we now know, that in the process the entire moral tradition of the West was overturned in an orgy of rationalized hatred, torture and killing.[4]

In this technological and moral revolution, the conquest of disease served as a grand paradigm for the Nazis. This was possible because by the early twentieth century, disease was no longer seen as some mysterious curse, but as the logical outcome of certain chemical and organismic processes. Further, a whole body of work in sociology and racial science was taking the diseased body of the individual as a microcosm of the body politic more generally. That is, just as an individual became sick when the body was infected by foreign germs or parasites, so too did a society or nation show signs of malfunctioning when it was

---

[3] www.twingroves.district96.k12.il.us/Renaissance/Globe/BlackDeathHist.html; accessed December 31, 2006.

[4] See my *Morality After Auschwitz: The Radical Challenge of the Nazi Ethic*.

overrun by incompatible populations and institutions. And just as health was restored to an individual when outside germs or parasites were eliminated, so, nationalist policy-makers suggested, could the social body be cured by the removal of pernicious foreign elements. This scientific approach to heal the national body, politic became especially urgent, and attractive, in Germany in the years following World War I. In fact the incorporation of eugenic, and subsequent genocide, programs into Nazi public policy can be understood as the practical applications of medical science, as they understood it, to the social problems of post World War I Germany.

The intellectual bridge allowing the Nazis to move from the individual to the social body was provided by 'Darwinism', or more precisely, social Darwinism. According to Darwin's *The Origin of Species*, life forms in the natural world were engaged in an endless battle for existence. Those individuals most fitted to their ecological niche, that is, most able to find shelter and nourishment while fending off predators, were more likely to survive and reproduce than their less adept relatives. Over generations, Darwin's law of natural selection asserted: those traits that gave certain individuals an advantage thus came to define a distinct species. Populations that failed to evolve in this way were doomed by the very laws of natural selection, to become marginalized or even fully extinct. Further, because nature itself was in a constant state of slow change, this battle to survive never ceased. It is important to stress at this point that for classical Darwinism, the notion of 'the fittest' thus had a very local reference, namely to those individuals of the species in question that were most fitted to the environment in which they found themselves at some particular point in time. There was no 'absolute' sense to fitness, because traits that were useful in one age might well prove to be irrelevant or even dysfunctional in another.

Social Darwinism, especially as appropriated by the Nazis, was based on a particular reading of this theory. For our purposes, two elements of Nazi reading of social Darwinism are of particular significance. One was to take the scheme designed by Darwin to explain plant and animal evolution over geological time and apply it to the immediate needs of a given society. According to this reading, racial groups (or 'nations' in the parlance of the nineteenth century) were understood to be engaged in a struggle for survival against other races (or nations) in a way completely analogous to what Darwin described as the process of natural selection in the natural environment. This carried with it the implication that conflict to the death rather than co-existence was the law of nature, and that it applies to political groupings, such as nations, just as it applies to species. The second reading was to take the category of 'the fittest' to be referring not just to whatever turned out to be more useful for survival in a certain time and place, but to be an objective and absolute standard. Thus in the Nazi case, Aryans were taken to be the most 'fit' species (or 'race') of humans. They alone were able to create advanced civilizations and so their survival had an absolute moral priority over all others.

From the Nazi perspective, these two elements of social Darwinism combined, indicated that the defeat of the Aryan race during World War I was not merely a German problem, but a human catastrophe of global scale. The question was what kind of strategy was needed to get 'the fittest' race back on its rightful track. Aryan policy makers found the answer, for the reasons noted below, in the medical profession. Herein lay the rational for the full implementation of eugenic policies so as to allow the rational course of human evolution to proceed.

In devising a strategy, the Nazis drew on two developments in the late nineteenth century. One was the emergence of 'racial science', or eugenics, which held that racial traits were carried in the blood. The person probably most responsible for popularizing this view was Count Jean Arthur de Gobineau, who died in 1882. He argued in his *Essai sur L'inégalité des Races Humaines* of 1855 that the Aryan race was the only bloodline that carried the potential for creating culture. Other races, he argued, were able at best only to borrow or imitate the culture of the Aryans. Although there was some disagreement as to which present-day group represented the 'real' superior races (the French proposed the French, the Russians the Slav and so forth), there was wide agreement among anthropologists that racial blood did in fact make a difference and that there was one in the end, one race, that was superior to all others. It was further taken to be the case that the innate civilizing genius of the superior race could be weakened by 'racial mixing'. This explanation of matters was exploited especially by the Nazis after the German defeat of World War I. The German Aryans were defeated, the Nazis asserted, not because of their innate inferiority but because of bad blood, that is, an infected gene pool. Matters could be set right simply by removing the impurities through a social policy of rationalized breeding. Biologist, physicians and other health professionals had precisely the means at hand and were accordingly recruited, and willingly served.

The second development was the result of a series of virtually simultaneous advances in the second half of the nineteenth century which provided a scientific framework for defining how such blood became diseased and so how one might go about purifying. The first of these discoveries was the finding by Louis Pasteur that at least some symptoms of disease could be caused by microorganisms. His breakthrough began with his proposal that fermentation was not just a chemical reaction, but was the result of the biological processes of tiny living organisms. This discovery had of course certain immediate practical applications in brewing and in stopping spoilage (through 'pasteurization'). But Pasteur pushed his discovery further by speculating that if micro-organistic 'germs' could cause fermentation, then maybe they might also be the cause of infections and contagious diseases. Although this was not an entirely new theory, Pasteur was able through a program of experiments scientifically to demonstrate for the first time that microorganisms were indeed responsible for a number of communicable diseases such as potato blight, anthrax and even rabies. His experiments with inoculations and vaccines (a process already known in connection with smallpox) established his theory in the popular mind. More importantly for the thesis being developed here, various natural phenomena, like fermentation and infection, were now reduced to comprehensible physical and chemical processes.

At roughly the same time that Pasteur was developing his germ theory, a London physician, John Snow, conducted what has become a classic epidemiological study. He was reacting to the outbreak of a cholera epidemic in certain neighborhoods of London in 1854. He was motivated by a desire to understand why the spread of the outbreak traced the pattern that it did. Two aspects of this outbreak drew Dr. Snow's particular attention. One was that it seemed limited to only a few select areas. The other was that it broke out in a new neighborhood, Soho, over the course, seemingly, of one day (August 31st). For Snow, steeped as he was in rationalist scientific theory, it was inconceivable that these data could be dismissed as mere vagaries of nature. He was sure there was a rational, mechanistic reason for why things happened just this way and he set out to figure out why. An account of his activity notes that,

from day one, he patrolled the district, interviewing the families of the victims. His research led him to a pump on the corner of Broad Street and Cambridge Street, at the epicenter of the epidemic. "I found," he wrote afterwards, "that nearly all the deaths had taken place within a short distance of the pump." In fact, in houses much nearer another pump, there had only been 10 deaths – and of those, five victims had always drunk the water from the Broad Street pump, and three were schoolchildren who had probably drunk from the pump on their way to school.

Dr Snow took a sample of water from the pump, and, on examining it under a microscope, found that it contained "white, flocculent particles." By 7 September, he was convinced that these were the source of infection, and he took his findings to the Board of Guardians of St James's Parish, in whose parish the pump fell.[5]

The work of Dr. Snow had obvious health implications, especially for the prevention of cholera, and eventually other epidemics. But it had a broader philosophical implication as well. His analysis showed that the spread of a disease through a population followed rational laws. Like Pasteur's germ theory as regards the individual, epidemiological studies showed that disease on a grander scale was also susceptible to human understanding, and control. Plagues, once the sign of divine displeasure, now became secular phenomena that could be dealt with by the application of certain rational, scientific procedures.

The third body of research, which established germs as the link between Pasteur's work on disease and the studies of Dr. Snow of how disease spread among human populations, was provided a few years later by Robert Koch. Koch's contribution was based on his isolation of the *anthrax bacillus* (1877) and his demonstration, by injecting the cultivated bacilli into a healthy animal, that the bacillus he had isolated was in fact the causal agent of the disease. He followed this breakthrough with the identification and isolation of the organism that causes tuberculosis (1882). Maybe just as significantly, he was able to demonstrate that the bacillus that caused anthrax produced 'spores' that could survive and travel through hostile environment, giving birth to the harmful bacillus only when implanted in a suitable environment. Following these breakthroughs, Koch was sent to Egypt in 1883 to study a cholera epidemic. There he successfully identified the microorganism that caused the disease, described its mode of distribution and helped establish public health protocols for controlling its spread. In so doing, Koch married the germ theory of disease to Snow's epidemiology, to produce a social policy for managing public health on a large scale. Along with Pasteur and Snow, Koch helped establish firmly in the public mind that well-being and disease were very much a part of the blood, that disease was the manifestation of the presence in the blood of hostile organisms, and that both environmental controls and individual strength were factors in determining one's health and survivability.

As the focus of public health shifted attention away from the individual and on to the group, many individual pathologies came to be understood as part of larger

---

[5] From Judith Summers count of The Broad Street Pump Outbreak, cited from http://www.ph.ucla.edu/epi/snow/broadstreetpump.html; accessed December 31, 2006.

social patterns. In an age obsessed with race and with national characteristics, it is not surprising that the blood of a 'folk' became the object of scientific interest and speculation. Racial theory, social Darwinism and eugenics were intermingled so as to produce a medical model for achieving political, economic and social improvements. As Robert Proctor puts it: "In the eyes of its founders (Alfred Ploetz and Wilhelm Schallmayer), racial hygiene was supposed to complement personal and social hygiene; racial hygiene would provide long-run preventive medicine for the German germ plasm."[6]

Social engineering, however, like the laws of Darwinian evolution, was devised with the survival of the species or nation in mind, not the fate of individuals. In fact, inferior individuals often had to die for the long-term benefit of the species. [7] Just as physicians routinely effected cures by exorcising harmful elements from the individual's body, so individual persons in a society might have to be eliminated to promote the health of the collective. Fascist regimes, such as that of the Nazis, certainly had no problem subordinating the interests of the individual to the needs of the collective.

Perhaps the year 1914 can serve as a symbol of this new paradigm. In that year one of the greatest engineering achievements of history reached its successful conclusion with the opening of the Panama Canal. The completion of the canal was testimony of course to the incredible advances made in engineering and construction technology. But the canal also reflects a medical triumph, namely the control of Malaria. This impressive victory over disease and nature took place only weeks after Austria declared war on Serbia, thus setting in motion the chain of events that led to World War I. Here too, engineering and the new discoveries of medicine played crucial roles. This war saw both the introduction of the armored tank and aircraft, and also the wide use of chemical agents to cause artificial epidemics of disease and death. The medical sciences had been harnessed, whether for good or evil, as never before.

---

[6] Robert N. Proctor, "Nazi Doctors, Racial Medicine, and Human Experimentation," in George Annas and Michael Grodin, eds., *The Nazi Doctors and the Nuremberg Code: Human Rights in Human Experimentation* (Oxford: Oxford University Press, 1992), p. 18.

[7] Konrad Lorenz, who won the Nobel prize in Medicine in 1973 is quoted as having said: "The individual is almost nothing and society is everything." Cited from: Benno Mueller-Hill "Eugenics," in Arthur Caplan, ed., *When Medicine went Mad* (Totowa, NJ: Humana Press, 1992), p. 45.

It is within this world of rapidly expanding medical knowledge, specifically the ideas that the health of a population, or race, was a function of the blood and that diseases were caused by foreign organism, that the Nazi policy of curing the social body of Aryan Germany took shape. Social Darwinism allowed the Nazis to see the advancement of the Aryan race as linked to, but transcending, the welfare of its component individuals. In light of this paradigm, a range of Nazi policies make sense. One was the establishment of an appropriate niche in which the German body could grow; hence the concept of *Lebensraum*. A second was the policy of building up the physical bodies of individual Germans through The Hitler Youth with its various sports and physical fitness programs. A third was a dense thicket of laws and policies designed to assure the quality of German blood through regulations on those marriage, the carrying out of sterilizations and other eugenic activities on those unworthy of procreation, the establishment of euthanasia programs for genetic misfits, and the purification of the population through migrations, transfers and/or exterminations of inferior populations.[8] In short, one could look at the larger Nazi program as a massive exercise in public health through which harmful organisms would be purged from the Aryan people and a nurturing ecology carved out, allowing the natural dominance of the German people to be realized. This orientation to the treatment of Jews, especially among physicians, is articulated well by Arthur Caplan in his summarization of the defense arguments in the *Doctors Trial* at Nuremberg, Caplan writes:

> Physicians justified their actions [...] on the grounds that the Jews, the homosexual, the congenitally handicapped, and the Slave posed a biological threat to the existence and welfare of the Reich. The appropriate response to such a threat was to eliminate it, just as a physician must eliminate a burst appendix using surgery or a dangerous bacterium using penicillin. Viewing specific ethnic groups and populations as threatening the health of the German state permitted, and in the view of those on trial demanded, the involvement of medicine in mass genocide. The overarching biomedical paradigm provided the theoretical

---

[8] It should be noted that the practice of eugenics along these lines was hardly limited to the Nazis, or even to Germany. In fact the United States, until the rise of the Nazis to power, was in the forefront of eugenics and sterilizations. See Robert N. Proctor, "Nazi Doctors, Racial Medicine, and Human Experimentation," in George Annas and Michael Grodin, eds., *The Nazi Doctors and the Nuremberg Code: Human Rights in Human Experimentation* (Oxford: Oxford University Press, 1992), p. 21. Other countries, like Russia and Brazil, were pursuing similar programs.

basis for allowing those sworn to the Hippocratic principle of nonmaleficence to kill in the name of the state.[9]

The array of laws, regulations and institutions brought into play by the Nazis were not seen as breaching the norms of natural law. To the contrary, they were understood to be the expression of the latest medical insights into the mechanisms of human destiny, and were run by young idealists, convinced that they were serving the larger good by working the levers of the natural world. Through the dispassionate application of racial hygienic policies, the 'new Germans' of the *Third Reich* were on the leading edge of what they regarded as the ultimate solution to the social problems of the West. They could even be seen as bringing to fulfillment the Bible's messianic promise that humankind would come to exercise divinely ordained rulership over nature.[10]

More broadly, Nazi genetic and medical policy reflected the modern confidence that nature was no longer innate and immutable, but was open to human manipulation and even betterment. Nature, as it was, could be tamed and domesticated. It is only in the wake of the Nazi genocide that the West began to rethink its own rationality and the extent to which humans should try to rule over nature. In the end, the human attempt to outdo nature in the creation of the perfect human life left a legacy of pain, suffering and death. This, as the Greeks taught us, is the ultimate punishment of *hubris*.

[9] Arthur Caplan, "The Doctors. Trial and Analogies to the Holocaust in Contemporary Bioethical Debates," in George Annas and Michael Grodin, *The Nazi Doctors and the Nuremberg Code: Human Rights in Human Experimentation* (Oxford: Oxford University Press, 1992), p. 268.

[10] See Christian Pross, "Nazi Doctors, German Medicine, and Historical Truth," in Annas and Grodin, *The Nazi Doctors and the Nuremberg Code: Human Rights in Human Experimentation* (Oxford: Oxford University Press, 1992), p. 32.

In Response to Peter Haas

Margaret Brearley

Peter Haas has written perceptively elsewhere on the Nazi 'healing/killing' paradox. His present thesis, that 'precisely the medical discoveries of the late nineteenth century [...] lay behind the Nazi attempt to create a new natural, and so social, order. [...] under which the Aryan race would flourish,' has many fascinating ramifications. One could extend his argument further, proposing, for example, that the work of Hippolyte Bernheim on susceptibility to hypnotism, *Suggestive Therapeutics* (1888), of Pierre Janet on dissociation in hystericals (1892) and of Morton Prince, *Dissociation of personality* (1906), may have provided valuable tools in the Nazi development of mass propaganda and the psychological training of personnel for mass killing.

Moreover, if Haas is correct, then it is deeply ironic that many of those medical discoveries were made by Jewish doctors (above all in Germany and Austria) – particularly in fields related to those 'human pathologies' which Haas emphasises as crucial to the Nazi world-view. Jacob Henle anticipated the germ theory of infection, while Julius Cohnheim proved that pus cells are derived from bloo. Jewish doctors had preeminence in bacteriology, immunology, hematology, histology and microscopic medicine. Jews became conspicuous in the discovery of bacteria and the development of immunologic methods for diagnosing and preventing bacterial infection; the test for typhoid fever and vaccines against cholera and plague, for example, were devised by Fernand Widal and Mordechai Waldemar Haffkine respectively. Other nineteenth-century Jewish physicians investigated infectious diseases and congenital diseases such as syphilis, while August von Wasserman introduced the first diagnostic test (1906) and Paul Ehrlich the first effective drug for syphilis (1910).[1] (Significantly, Franz Boas, a Jewish anthropologist, demonstrated conclusively in *The Mind of Primitive Man* (1910), a book later burned by the Nazis, that race theory was a fallacy).

Jewish physicians and medical scientists thus contributed uniquely and disproportionately to the health of German-speaking peoples. Paradoxically, beneficent Jewish science was attacked by Otto Weininger and others as materialistic, as means to an end, lacking in transcendence: "the Aryan experiences the attempt to

---

[1] *Encyclopedia Judaica* (Jerusalem: Keter, 1972), vol. 11, pp. 1196-98.

understand everything [...] as a devaluation of the world, for he feels, that it is precisely the unfathomable, which gives existence its value. The Jew has no fear of mysteries, for he perceives none."[2]

Such linkage between Jews and disease had occurred during the Black Death in 1348. Haas view that the Black Death was attributed solely to 'the natural order' and 'punishment from God', needs amplifying. For already in 1348 it was blamed on contemporary Jews, accused of an international conspiracy to poison Christendom.[3] Jews in Savoy 'confessed' under torture to well-poisoning. As a result, many thousands of Jews were slaughtered and burnt alive in over two hundred communities, in France, Alsace and especially in Germany, sometimes with episcopal blessing and often by the same penitent flagellants whom Haas mentions. The idea – enhanced by the legendary skill of Jewish physicians – that Jews bring sickness and death to Christians, deliberately contaminating them through doctored wine or meat, had circulated in Europe at least since the ninth century.[4] The Black Death simply intensified the notion, later developed by Nazism, that Jews contaminate gentiles with disease and that Jews must therefore be murdered.

A related notion, pre-dating the Black Death, that Jews desire gentile blood for their own healing, derived from the widespread superstitious use by Christians of human parts, especially blood, for medicinal and magical purposes.[5] Worse, Jews were accused, as in Freiburg in 1401, of using the dried blood of murdered Christian children to cause plague affecting men and cattle.[6] From blood libel myths it was only a small step to the accusation that Jews poisoned Christian blood itself and ultimately to the Nazi attribution of defeat in World War I, as Haas suggests, to 'bad blood [...] an infected gene pool'.

Nazism did indeed portray itself as summoned to cleanse that gene pool, as 'a massive exercise in public health through which harmful organisms would be purged from the Aryan people'. Haas cites no leading Nazis to support his con-

---

[2] Otto Weininger, *Geschlecht und Charakter*, 26th. ed. (Vienna/Leipzig: Wilhelm Braumueller, 1925: 1st pubd. 1903), p. 417; Translation from German by the author.

[3] Jacob R. Marcus, *The Jew in the Medieval World* (New York, NJ/Philadelphia, PA: Meridian Books/The Jewish Publication Society of America, 1960), pp. 43-48.

[4] Joshua Trachtenberg, *The Devil and the Jews: The Medieval Conception of the Jew and its Relation to Modern Anti-semitism* (Philadelphia, PA/Jerusalem: Jewish Publication Society of America, 1983), pp. 97-100.

[5] Trachtenberg, *The Devil and the Jews*, p. 143, p. 148-151.

[6] Trachtenberg, *The Devil and the Jews*, p. 144.

tention, but Goebbels referred to 'a life-and-death struggle between the Aryan race and the Jewish bacillus.' Hitler himself repeatedly referred to Jews as bacilli and stated, in February 1942: "The Jew will be identified. The same battle that Pasteur and Koch had to fight must be led by us today. Innumerable sicknesses have their origin in one bacillus: the Jew!'[...] We will get well when we eliminate the Jews."[7]

But while Haas rightly emphasises the role in Nazism of rationality in applying the consequences of Social Darwinism and medical advances, he perhaps underestimates the profound role of ideology and, especially, of irrationality. For although Hitler claimed, contrary to all the evidence, that "pure and applied science [was] an almost exclusively Aryan achievement,"[8] nevertheless he argued that not open but arcane, occult knowledge was the key to controlling nature: "[Only] when knowledge re-acquires the character of secret, initiate knowledge, and ceases to be accessible to all and sundry, will it again fulfil its normal function, namely that of being the means and the power to control both human and non-human nature."[9]

Hitler's essentially irrational, anti-scientific view is especially significant since he was widely viewed as not only Germany's saviour, but also as its physician. Willy Heidinger, managing director of DEHOMAG, which ran the early IBM computing machine, described Hitler in 1934 as "our nation's Physician", whom DEHOMAG was "proud to assist" with information on every single member of the nation, "so that our Physician can take corrective procedures to correct the sick circumstances [...] We have the deepest trust in our Physician and will follow his instructions in blind faith."[10]

Hitler implicitly adopted the role of physician in his unpublished *Second Book,* attacking 'the medicines of our body politic' (emigration and a reduced birth rate) prescribed by pacifist economists and Marxists. His own "prescription" for "healing the body politic of a profound and serious sickness" included counter-

---

[7] Cited in Peter Longerich, *The Unwritten Order: Hitler's Role in the Final Solution* (Stroud/ Gloucestershire: Tempus, 2001, 2003²), p. 173, p. 156.

[8] Adolf Hitler, *Mein Kampf*, trans. Ralph Manheim (London: Hutchinson,1974), p. 317.

[9] Hermann Rauschning, *Gespräche mit Hitler*, (Zurich/ New York: Europa Verlag, 1940), p. 40; Translation from German by the author.

[10] Cited in Edwin Black, *IBM and the Holocaust* (New York, NJ: Three Rivers Press, 2001), p. 51.

acting "one poison with another."[11] Hitler, as supreme physician, was driven, as he frequently stated in *Mein Kampf*, by one single, dominant idea: "the holiest obligation [...] to see that the blood is preserved pure" to ensure the "regeneration, which gradually eliminates racial poisonings."[12] He used rational, scientific language to cloak and make palatable the profound irrationality of his programme. He used medical terms to camouflage "the eradication of all of European Jewry" as "biological," resulting in the "clean, organised body of the *Volk*."[13] Heydrich described mass murder by the *Einsatzgruppen* in Eastern Europe as 'self-cleansing efforts' – *Selbstreinigungsbestrebungen*[14] – as though they were – in Hitler's euphemistic phrase – Nature's "corrective decisions."[15] One euphemism used by both Nazis and 'race hygiene' – scientists for killing was *Aufartung durch Ausmerzung*, essentially "physical regeneration through eradication," while another was "decontamination."[16]

The medical camouflage given to mass murder, both in the T4 euthanasia programme and the Final Solution, was further enhanced by "Hitler's order that only physicians should kill."[17] Hitler stated to Goebbels in February 1942 that the "annihilation" of Jews would be accelerated "with cold ruthlessness [...] we shall be rendering an inestimable service to suffering humanity that has been tortured by Jewry for thousands of years."[18] Future genocide would simply be the healing of infection caused by the Jews themselves. In Alfred Rosenberg's words: "where any kind of wound is torn open in the body of a nation, the Jewish demon always eats itself into the infected part."[19]

By masking pitiless inhumanity with medicalised language, Rosenberg portrayed murder as medically and morally necessary: "That which is alien must be unflinchingly excised, or if necessary, destroyed", and Nazism itself as irresistible

---

[11] Gerhard L. Weinberg, ed., *Hitler's Second Book: The Unpublished Sequel to Mein Kampf by Adolf Hitler*, trans. Krista Smith (New York, NJ: Enigma Books, 2003), p. 108, p. 44.

[12] Hitler, *Mein Kampf*, p. 365, p.364.

[13] Longerich, *The Unwritten Order*, p. 149, p. 212.

[14] Longerich, *The Unwritten Order*, p. 109.

[15] Hitler, *Mein Kampf*, p. 363, p. 364.

[16] Henry Friedlander, *The Origins of the Nazi Genocide: From Euthanasia to the Final Solution* (Chapel hill, CA/London: University of North Carolina Press, 1995), p. 20, p. 231.

[17] Friedlander, *The Origins of the Nazi Genocide*, p. 219.

[18] Longerich, *The Unwritten Order*, p. 157.

[19] Alfred Rosenberg, *The Myth of the Twentieth Century* (Sussex: Historical Review Press, 2004), p. 298.

"present-day avenging Nature."[20] Since "the sun myth of the Aryan is not only transcendental but also a universal law of nature and biology", the irrational pseudo-religion of Nazism could be presented as holy: "today a new faith is awakening - the Myth of the blood; the faith that to defend the blood is to defend the divine nature of man", and the swastika, symbol of the sun, "of fertile, ascending life", as a source of healing.[21]

---

[20] Ibid., p. 68, p. 102.
[21] Ibid., p. 82, p. 65, p. 99.

In Response to Peter J. Haas

John K. Roth

> I am the Lord your God, who brought you out of the land of Egypt, out of the house of slavery; you shall have no other gods before me.
>
> Exodus 20:2

For decades, Peter Haas has shed an important light on the links between Nazi Germany's genocide against the Jewish people and the modern developments in science, philosophy, and religion, that provided the seedbed for that catastrophe. Haas' essay on "*The Medicalization of Nature*" continues that distinguished output, but its retrospective analysis leaves urgent questions in the lurch.

Haas' historical account is on target. He shows how the Nazis' intermingling of racial theory, social Darwinism, eugenics, and expanding medical knowledge contributed to the Holocaust. Far from sensing that they were violating natural law or biblical religion, Nazi ideologues – scientists and physicians among them – saw themselves as agents of political, economic, and social progress. In their view, the elimination of 'inferior' racial groups, especially Jews, and 'unproductive' members of society, became, in Haas' words, "a massive exercise in public health through which harmful organisms would be purged from the Aryan people, and a nurturing ecology carved out, allowing the natural dominance of the German people to be realized."

The first question that Haas leaves in the lurch emerges from his historical account and the fact that the Holocaust did not have to happen. It resulted not from necessity or fate, but because people made choices and decisions. Those choices and decisions, of course, were made in social, political, religious, and scientific contexts, but the contexts alone do not explain, at least not completely, what individuals and groups decided to do. Haas does not overlook these realities, for his essay ends by emphasizing a theme that threads throughout it, namely that *hubris* contributed mightily to the Third Reich's mass atrocities. Pride has long been called one of the 'seven deadly sins,' but why did hubris loom so large in the Nazi project? A recitation of historical events, even an analysis of them, may beg that question more than answer it. If pride drove the Nazi project, then pride's impact cannot be explained, at least not entirely, solely by appeal to that project and its conditions. Attempts to do so would be circular. Inquiry about the sources and powers of pride entails psychological, philosophical, and religious dimensions that historical analysis alone cannot provide. Perhaps no analysis can do so, but Haas would help his readers by plunging more thoroughly into inquiry about pride's sources, dynamics, and dilemmas that his essay definitely brings to mind.

The importance of Haas' dealing with the sources and powers of pride is not restricted to his interpretation of the Nazis' medicalization of nature. Readers of his essay are likely to discern another key thread in his account, one that has loose-end qualities because Haas leaves the strand more implicit than explicit. The Nazi project is over, its aftereffects are not. They may include an outlook of the following kind: Nazi science and medicine went badly wrong, but the disastrous errors of their ways do not impugn the immense progress and promise of science, medicine, and technology. Today and tomorrow, we should and will avoid the policies and practices, even the pride, that made science, medicine, and technology Hitler's allies.

Haas' readers will rightly sense that he would never buy such lines of thought, but he needs to tell us *why*, an accounting that is likely to make him grapple with pride, among other things, as a force not confined to the past but perniciously present in contemporary life as well. More emphasis on this matter would also enable Haas to do the helpful work of identifying the issues and developments about which we need to be warned and corrected as far as human overreaching in the twenty-first century is concerned.

As Haas ponders the questions his essay leaves in the lurch, religion is likely to loom even larger than his discussion about the medicalization of nature suggests. I make this point because his essay makes me think of 'Deadly Medicine,' the special exhibit mounted at the United States Holocaust Memorial Museum in 2004-2005[1]. During my visits to the 'Deadly Medicine' exhibit, a particular document riveted my attention. The exhibit identified its source as a treatise issued by a Berlin publisher in 1935. Its author, Dr. Hermann Böhm, called his work *Darf ich meine Base heiraten?* (May I Marry my Cousin?).

Before noting the content of the document from this monograph about marriage, some glimpses of its author are instructive.[2] His fate is unknown to me, but

---

[1] The exhibit was accompanied by a helpful book. See Dieter Kuntz, ed., *Deadly Medicine: Creating the Master Race* (Washington, DC: United States Holocaust Memorial Museum, 2004). For the document that I discuss below, see *Deadly Medicine*, p. 72.

[2] The discussion about Hermann Böhm draws on my book *Ethics during and after the Holocaust: In the Shadow of Birkenau* (New York, NJ: Palgrave Macmillan, 2005), pp.

Böhm, who was born in 1884, was a German university professor and research institute director who specialized in genetics and racial science. In 1936, shortly after Böhm's report appeared, Dr. Gerhard Wagner, founder of the National Socialist German Physicians' League in 1929 and subsequently the Reich Physicians' Führer (*Reichsärzteführer*), tasked Böhm, his fellow Nazi, with an important educational assignment: he would teach National Socialist views on genetics and race to German physicians. Specifically, he would do so at an important training center, the *SS Doctors' Führer School*, an arm of the Physicians' League. The school's six-week courses supplemented traditional medical training by "character building" activities –"manual, mental, and moral," to use Robert Proctor's words – whose purpose was to impress upon young and promising doctors, nurses, and midwives their importance as leaders in promoting the Third Reich's program of racial hygiene.[3] Under Böhm's tutelage, the emphasis was less on curing illnesses among the living than on encouraging preventive measures to protect the health and purity of 'German genetic streams.' At the Doctors' School, Böhm created an Institute for Genetics. Its library and laboratory facilities advanced teaching and its research put German racial purity at the forefront of a Nazi blending of science, medicine, and politics that reflected and extended a resolutely antisemitic and ultimately genocidal world view.

Returning to Böhm's account about marriage, its best-known part was called *Zehn Gebote für die Gattenwahl* (Ten Commandments for Choosing a Mate). A prefatory paragraph indicates that the commandments were promulgated by "The Reich Committee for the National Health Service, together with the Reich Ministry of the Interior, the Reich Office of Health, and the NSDAP Office for Racial Policy." Böhm's treatise not only indicated his support for those agencies but also added his scientific prestige to their credibility. Meanwhile, the commandments were widely distributed in an attractive poster format by the Reich Committee for Public Health.

These *Zehn Gebote* had an imperative ethical content. "You should keep your mind and spirit pure!," "Marry only for love," "You should wish for as many children as possible." But what is particularly interesting and disturbing is the

---

85-6. My biographical sketch of Böhm relies on Robert N. Proctor, *Racial Hygiene under the Nazis* (Cambridge, MA: Harvard University Press, 1988), pp. 83-5. For this reference and other information about the ten commandments regarding marriage in Nazi Germany, I am indebted to Susan Bachrach at the United States Holocaust Memorial Museum. She served as the project director for the Museum's 'Deadly Medicine' exhibit.

[3] See Proctor, *Racial Hygiene*, p. 83.

way in which this document drew upon traditional religious forms and categories to promote policies and to legitimate powers whose 'logic' eventually led to Birkenau.[4]

Two instances, obvious ones, illustrate what I mean. First, Böhm's commandments were intended to be a kind of Decalogue; they took the form of the biblical Ten Commandments. That form was used to give them authority, to enhance the normative power of their content. Second, there is a striking contrast between the biblical Decalogue and the one that Böhm supported. It might not have been fully conscious to German readers in 1935, and it might be even less so today, but there are still resonances and echoes that deserve comment. The first commandment in Böhm's list is "Remember that you are a German." At least as the Hebrew text is often construed, the first commandment in the biblical Decalogue is "I am the Lord your God, who brought you out of the land of Egypt, out of the house of slavery; you shall have no other gods before me" (Exodus 20:2). Böhm, I am confident, knew that his commandments drew on a biblical form. Whether he sensed any dissonance between his first commandment and the one that Moses communicated to his people, is a harder call to make, but one worth pondering when questions about pride are raised. Böhm and his Nazi comrades used ethics and religion to promote pride, antisemitism, racism, and eventually genocide – all of them intertwined. What would have to happen for ethics and religion to be effective antidotes for such tendencies? Haas has much of value to say about such matters, and he needs to share his thinking about them more fully in the context of his reflections on the medicalization of nature.

While thinking about the history discussed by Haas and the other contributors to this volume about the Holocaust and Nature, my reflection often turned to a book called *How Green Were the Nazis?*[5] It concentrates not on medicine and medicalization, but on environmental concerns and objectives in the Third Reich – those among them concerning the German landscape – especially before World War II began in September 1939, ravaging the European continent and its populations.

The book's cover uses a haunting aerial photograph that depicts a German pine forest in Brandenburg. Formed by larch trees, the deciduous conifers that turn bright yellow in the autumn before shedding their needle-like leaves, a large

---

[4] See Kuntz, ed., *Deadly Medicine*, p. 72.

[5] See Franz-Josef Brüggemeier, Mark Cioc, and Thomas Zeller, eds., *How Green Were the Nazis?: Nature, Environment, and Nation in the Third Reich* (Athens, OH: Ohio University Press, 2005).

swastika stands out amidst the evergreens. The editors' note about the cover explains: "Presumably a zealous forester planted the sylvan swastika, with a diameter of nearly two hundred feet, in the 1930s as a sign of his allegiance to the Nazi regime. The swastika survived not only World War II, but also four decades of East German communist rule. Visible only from the air during the fall and winter months, the larch trees were detected in 1992 and felled in 2000."[6] Recently I planted a larch tree at my home in the Cascade mountains of north central Washington state. Doing so was a kind of protest against Nazi pride that found it good to use those tall, thin, magnificent specimens to mark nature itself with the swastika. No tree deserved that fate. As I look at the cover photo of *How Green Were the Nazis?* while writing these words about Haas' essay, I wonder: 'How proud were the Nazis – and why? How proud are we – and to what ends?' I hope that Haas will help his readers to know more and better about how to wrestle with those questions.

---

[6] Ibid., p. iv.

In Reply to My Respondents

Peter J. Haas

Margaret Brearley has added important language that further illustrates my point that the use of scientific language, and the language of medicine in particular, was integral to Nazi thinking and discourse. I thank her for these. She also very correctly points out that the notion of blood as a source of both disease and healing has a long history, going back at least as far as the Black Death. One can see this theme operating already in the aftermath of the Spanish Reconquista and the expulsion of Jews from Spain and Later Portugal. The 'cleanness' or 'purity of blood' (limpieza de sangre) became a social tool allowing for discrimination against Conversos (Jews converted to Christianity) or Moriscos (Muslims converted to Christianity) by established, 'old' Christian commoners. In either case, the connection between Jews, blood, and healthiness became a fixture in Western thinking that made the Nazi use of it much less revolutionary than it might at first appear.

Brearley raises two other points that require further amplification. One is the observation that Jews constituted a disproportionately large portion or the contributors to the medical (and other scientific) advancements of Germany and the West. As she notes, it is thus a peculiar irony that it was precisely the Jews who became victims of these advances. But this very fact hides another phenomenon that is one of the enduring questions of the Holocaust, namely the apparently widespread support Nazi ideology enjoyed among the intellectual elite, the friends and peers of a considerable number of the very Jews who were slated for destruction. It is an enduring point of embarrassment for the German medical profession that so many of its members acquiesced, if not actively and willingly participated, in Nazi 'health' initiatives such as the T-4 Euthanasia program and later medical experimentations on camp inmates. One would have thought that members of the medical profession in particular would be able to perceive the Nazi misappropriation of their science. To be sure physicians and other medical professionals were not the only ones to be absorbed into the Nazi program. Jurists, academicians and even Protestant theologians willingly involved themselves in the Nazi racial program. But the health profession with its significant Jewish component and which was in a position both to understand Nazi 'racial science' and to protest its false premises constitute a particularly troubling phenomenon.

There are of course several ways to account for this professional and moral abdication. One is the possibility that medical professionals felt empowered by the Nazi focus on medical expertise and so were drawn in as a matter of professional pride. Another possibility is that many may have felt the program of national regeneration embarked on by the Nazis was for them a call to a duty that transcended the mundane work of the clinic. There were also, social and professional pressures to conform to what was already going on, and of course along with that was sheer career opportunism. But a further element surely is another important strand to which Brearley points us, namely the mythic character of Nazi thinking. The complex and potent intermixture of science, technology and myth is a substantial topic on its own, one we see being played out again today in certain Islamicist Jihadi movements, for example. What the mythic component allows is for participants to interpret and deploy science and technology to certain transcendent ends that serve larger purposes of good and evil. It establishes a taxonomy of moral considerations that go beyond, and even trump, normal human feelings of empathy for the individual. The extent to which a Teutonic/Aryan mythology permeated Germany before and during the Nazi period is a fascinating story, and one to complicated to go into here. But we certainly know that a new mythology was being forged about the necessary ('natural') dominance of the Aryan peoples, and of course with that a need to rethink the whole role of Christian ethics of compassion, especially given its Jewish roots. It seems clear that in many of its facets, Nazism was the playing out on the terrestrial stage of a great cosmic battle.

Two points that Brearley raises are thus crucial for understanding the Nazi manipulation of nature, especially as regards public health and medicine. One is that the notion of blood as central to the character of not only a person but of a people, 'volk', religion and culture has a long history in the West. This made the Nazi appropriation of 'blood' language much less alien and more familiar than would otherwise have been the case. The second point is that importance of the Nazi world-myth of the destined superiority of the Aryan people, a myth that seems to have motivated many professionals to abandon their received ethics in order to participate in the creation of literally a new world order. It was of course this mythic transvaluation of values that turned modern advances in medicine and technology against the Jews, the very community that had contributed so much to these developments.

These thoughts about the power of myth turn me to John K. Roth's very important question about pride. I purposely ended my essay with the word 'hubris' rather than pride for a very specific reason. For me, 'hubris' carries a connotation that takes us well beyond pride *simpliciter*. Hubris indicates a sort of arrogance of pride, an oversized sense that one can accomplish more than would normally be possible or even thinkable. I think that Germans in the 1920's and 1930's have ample reason to be proud of their country's scientific and technological advancements on all sorts of fronts. As even Roth agrees, pride in and of itself is not pernicious. It becomes pernicious, however, when it moves one into arrogance and, to use Roth's word, 'overreaching'. Maybe the issue here is a matter of scale. It is one thing to be proud of one's ability as a heart surgeon, for example. One might even feel that one is so good as to be able even to train others to be excellent in the field. This may lead to a certain arrogance in the classroom and insufferability among colleagues, but these are, as it were, annoyances. It is another matter entirely to regard oneself as so good as to be able to make infallible life-and-death decisions on behalf of whole populations of people. That is, pride turns into hubris when one starts to act more cosmically, as if the normal laws of morality or of nature no longer apply, as if one had a sort of divine perspective that was beyond normal human evaluation.

It is this level of pride that I think came to characterize the Nazi regime. It was the sense that through the proper deployment of its scientific, technological and intellectual resources that Germany could not only improve the lot of Germans, but could in fact remake the very structure of nature on the human, animal, and geographical levels. It was the sense that the Nazi party, through its Leader, could now stand above time and space and determine the future course of human history. Roth gives us two striking examples of this, one in the 'ten commandments' issued by Dr. Hermann Böhm, which Roth rightly juxtaposes to the Biblical ten commandments, and the other the swastika planted in the Brandenburg forest. Neither of these examples is worrisome on its own. But sense as part of a larger all-encompassing program, they offer insight into the kind of hubris that grabbed the Nazi imagination. In these two simple acts we see reflections of an attempt not only to reshape the face of nature, but to displace the divine itself. It is at this level that we enter the cosmic mythic realm I mentioned earlier.

This Nazi pride in its own cultural prowess is hardly unique in human history. The word I used, 'hubris' is found among the oldest literature of the Western world, the epics of Homer. Roth is certainly right in warning us of the dangers of hubris in our own day. We learn lessons all the time about the limits of our abilities, whether it is flood-prevention dikes that fail, undersea oil wells that leak uncontrollably, or limited military engagements which never end. To be sure none of these reach anything near the level of ‘hubris’, of overweening and over-reaching pride, that we witnessed under the Nazis. But if we let pride in our accomplishments become part of a mythic structure of super-human dominance, if we seek to displace the moral teachings of our religious heritages with a newly created secular narrative of good and evil, then the Nazi experience should start to loom as an example of how badly things can go.

"They Say of Them, Let the Sacrificers of Men Kiss Calves"

(Hosea, 13.2)

Margaret Brearley

Max Horkheimer described National Socialism as "rebellion of Nature against civilisation,"[1] while Robert Pois dedicated an entire book to National Socialism and the religion of nature.[2] The early actions of the National Socialist government following their seizure of power on 30 January 1933, indicate the crucial importance of nature in their legislative programme. Laws were swiftly passed to ensure protection of the natural environment, far surpassing that already in place under the Weimar government.

Three laws concerning the protection of animals were passed in quick succession. The first, enacted in April 1933, forbade kosher slaughter of animals and had clear anti-semitic intent.[3] In May 1933 a law was enacted, intensifying penalties for cruelty to animals, followed on 24 November 1933 by a law closely regulating vivisection and forbidding cruelty to animals. Contemporary commentators noted that this unprecedented degree of protection was offered not on grounds of human feelings but, for the first time, for the sake of the animal it-

---

[1] Rolf Wiggershaus, *Die Frankfurter Schule* (Munich: Hanser, 1986), p. 389. Cited in Joachim Radkau, "Naturschutz und Nationalismus - wo ist das Problem?" in Joachim Radkau and Frank Uekoetter eds., *Naturschutz und Nationalsozialismus*(Campus, Frankfurt am Main, 2003), 40-54, p. 43.

[2] Robert Pois, *National Socialism and the Religion of Nature* (London/Sydney: Croom Helm, 1986), p. 11: "Hitler, Heinrich Himmler, Alfred Rosenberg, Robert Wagner, Ernst Krieck, Paul Josef Goebbels and others…viewed themselves as the bearers of the new religion of nature", p. 29.

[3] Edeltraud Klueting, "Die gesetzlichen Regelungen der nationalsozialistischen Reichsregierung für den Tierschutz, den Naturschutz und den Umweltschutz," in Radkau/Uekoetter, *Naturschutz*, pp. 77-105. The question of kosher slaughter, debated in the Reichstag since 1886, had been previously always decided in favour of religious freedom. pp. 79-80.

self.[4] In 1934 Hermann Göring enacted a *Reich Game Law*, making him the first *Reichsjaegermeister*, and made killing an eagle a capital offence.[5]
Extensive forest legislation, controlled from December 1933 by Göring himself, was enacted from January 1934 essentially in the interests of hunting and economic exploitation rather than of forest conservation.[6] In 1935 the *Reichsnaturschutzgesetz* was enacted to ensure comprehensive protection of nature throughout the Reich, in part to counteract the environmental destruction which had occurred since 30 January 1933, due to the prioritisation of economic development and mass employment programmes.[7] Significantly, it provided for confiscation without compensation on the National Socialist principle of *Gemeinnutz vor Eigennutz* (for the benefit of all rather than of the individual).[8]
The requirements of *Realpolitik* could, moreover, counteract the ideal of nature preservation. From 1934, state forests were required to double their income annually, as it were privately owned forests from 1936. By November 1937, forest conservation was nullified by intensified economic and later military exploitation. For reasons both of aesthetics and military strategy, new large-scale industrial works were built typically, concealed within forests.[9]
Much heathland and moorland was turned over to agriculture, in line with Hitler's dictum that no square metre of German ground should be left untilled.[10] Yet, despite "the basic premise of Nazism that farmers constituted the incorruptible nucleus of the *Volk*"[11] and early plans of party ideologue Gottfried Feder for rural settlement, the actual effect of National Socialist policies was a massive exodus from the land of an estimated 1,000,000 to 1,500,000 farm workers and their dependants by 1939.[12] After the invasion of Poland, Polish farmers were

---

[4] Klueting, "Die gesetzlichen Regelungen der nationalsozialistischen Reichsregierung für den Tierschutz, den Naturschutz und den Umweltschutz," p. 86.
[5] Simon Schama, *Landscape and Memory* (London: Fontana Press, 1996), p. 68.
[6] Klueting, "Die gesetzlichen Regelungen der nationalsozialistischen Reichsregierung für den Tierschutz, den Naturschutz und den Umweltschutz," pp. 88-9.
[7] Hansjoerg Küster, "Der Staat als Herr über die Natur und ihre Erforscher" in Radkau/Uekoetter, *Naturschutz*, 55-64, p. 57.
[8] Jürgen Trittin, "Naturschutz und Nationalsozialismus - Erblast fuer den Naturschutz im demoktischen Rechtsstaat?" in Radkau/Uekoetter, *Naturschutz*, 33-39, p. 34.
[9] Küster, "Der Staat," p. 56. Autobahns and Alpine high tension electricity cables were similarly concealed where possible.
[10] Radkau, 'Naturschutz' in Radkau/Uekoetter, *Naturschutz*, p. 45.
[11] Richard Grunberger, *A Social History of the Third Reich* (London: Penguin Books, 1974; 1st. pubd. 1971), p. 200.
[12] Grunberger, *A Social History of the Third Reich*, p. 207.

deported from certain areas to make way for German settlers.[13] Following Operation Barbarossa against the Soviet Union in June 1941, the great Bialowieza forest was cleared of its human inhabitants to safeguard its primeval wildness and enhance its hunting potential.[14]

Heinrich Himmler's *Reich Commission for Strengthening Germandom* included a planning agency committed to "total reshaping of large parts of the Polish territory into German ideal landscapes," destined for future German settlement.[15] Key academic landscape planners became senior SS officers within Himmler's agency. They included Erhard Maeding, the officer for landscape formation on Himmler's planning board, who envisaged that Germans would be "the first occidental people to form their own spiritual environment in the landscape," both restoring the "harmony of all living things," and creating a communal "space that is cleansed of the alien."[16] Some conservationists urged that the German landscape should be cleansed of roadside advertisements, interpreted as the intrusion of alien American-Jewish concepts of life and as a "pathogenic force for decomposition."[17] Walther Schönichen argued in his 1933 essay: "The German *Volk* must be cleansed – and the German Landscape?", that the landscape must be thoroughly cleansed of advertisements in order to maintain its role in "forming and maintaining the soul of the German *Volk*."[18]

Alien plants, too, were to be excluded where possible. From 1936, maps were made of 'natural vegetation', creating a plant sociology to allow for a 'correct' nature to emerge. The concept of 'the potentialities of natural vegetation' – flora indigenous to each particular region or locality if human intrusion were abandoned – was introduced in the planting of newly landscaped autobahns, with non-native plants, shrubs and trees excluded.[19] Similar notions were developed for gardens. Willy Lange, a leading academic on garden design, adapted the

---

[13] Joachim Wolschke-Bulmahn and Gert Groening, "The National Socialist Garden and Landscape Ideal" in Richard A. Etlin, ed., *Art, Culture and Media under the Third Reich* (Chicago, IL/London: Univ. of Chicago Press, 2002), 73-97, p. 90.

[14] Schama, *Landscape*, pp. 70-2.

[15] Wolschke-Bulmahn/Groening, "The NS Garden", pp. 80ff.

[16] Wolschke-Bulmahn/Groening, "The NS Garden", p. 83, p. 90.

[17] Friedemann Schmoll, "Die Verteidigung organischer Ordnungen: Naturschutz und Antisemitismus zwischen Kaiserreich und Nationalsozialismus," in Radkau/Uekoetter, *Naturschutz*, 169-182, p. 178.

[18] Walther Schoenichen, "'Das deutsche Volk muß gereinigt werden' - Und die deutsche Landschaft?," in *Naturschutz*, 14 (1932-3), 205-9; cited in Schmoll, "Die Verteidigung," p. 179; Translation from German by the author.

[19] Hansjoerg Küster, "Der Staat", pp. 58-60.

'Blood and Soil' ideology to gardens, seeing them as expressing the soul of a race. Rejecting formal gardens as expressing the 'race morass of the South', he argued for nature gardens and the use of native plants. Alwin Seifert, an influential garden designer, applied his concept of *Bodenständigkeit* (rootedness in the soil) to gardens, while garden architect Albert Kraemer anticipated 'race-specific' gardens originating in 'blood and soil''.[20]

The ideological link between landscape and 'race soul' contributed to the drive to enable more Germans to enjoy the German countryside. The *Kraft durch Freude* (Strength through Joy) organisation encouraged mainly internal tourism and, in the year prior to the outbreak of war, it was used by 8.5 million people. Campaigns to bring "Sun and Greenness to all Workers" promoted the beautifying of villages and factories, the proliferation of flowers, the creation of new public gardens. Yet, as Peter Reichel has noted, this apparent celebration of nature, and the greater access to it, provided by the cheap *Volkswagen* and the autobahns with their 9000 bridges, also served longer-terms goals of "war and destruction [...] the mastery of technology over nature."[21]

Indeed, despite the widely-propagated conviction deriving from German Romanticism and formulated by Heinrich Friedrich Wiepking-Jürgensmann, landscape advisor of Himmler and professor at the Berlin Institute for Landscape Design, that the German person is a "being close to nature" and that "the love for plants and the landscape bursts forth from our blood,"[22] National Socialism in fact wrought "comprehensive destruction on nature through improvement works, autobahn construction, intensification of forestry, the construction of industrial and military sites etc."[23]

Moreover, as Reichel has stressed, nature was functionalised as a tool in National Socialist propaganda. Innovative means were used to suggest that National Socialism itself was an unstoppable force of nature. Hitler was widely portrayed as the sun, and his flights were stylised as momentous natural events: his aeroplane was "a giant bird", its "touching the earth" aroused "a unique symphony of enthusiasm [...] the most powerful and elevating thing that Germany has ever

---

[20] Wolschke-Bulmahn/Groening, "The NS Garden", pp. 76-8.

[21] Peter Reichel, *Der schöne Schein des Dritten Reiches: Faszination und Gewalt des Faschismus* (Munich/Vienna: Carl Hanser Verlag, 1991), p. 246, p. 237, p. 283; Translation from German by the author.

[22] Cited in Wolschke-Bulmahn/Groening, "The NS Garden," p. 84.

[23] Klueting, "Die gesetzlichen Regelungen", p. 105; Translation from German by the author.

seen and experienced."[24] Natural calendar events were re-interpreted: 'May-Day' became the festival of Germany as spring, the harvest festival the day of Germany's blood.[25]

Nature imagery was used repetitively in propaganda to affect the reader's or listener's emotions and to hinder rational thought. Uriel Tal perceptively noted that sun, darkness, seasons, elements of weather, trees and so on were repeatedly employed to trigger unconscious responses: "Motives borrowed from nature [...] were converted into tangible, living entities in a form similar to animism."[26]

The clear dichotomy noted above between ideal love of nature and the actual destruction of it is not, in fact, incongruous. In National Socialist ideology, passivity towards Nature and rational, scientific analysis of it were both regarded as symptoms of non-Aryan or Jewish decadence. Radical intervention in nature was a fundamental Nazi tenet. Richard Walther Darre, chief architect of the 'Blood and Soil' ideology, argued that the inner nature of the Northern (i.e. Aryan) race as *Tatmensch* (man of deeds) enabled it to realise itself against Nature, indeed to actively overcome nature, rather than passively submit to it.[27] Moreover, Hitler held a conviction that power brought the right to booty: "Hitler believed it his right as conqueror to claim artworks as the spoils of victory."[28] This legitimated not only the seizure of art treasures on a scale unprecedented in history, but also the radical exploitation of natural resources such as granite, using, if necessary, slave labour.

Objective nature was of little importance compared to its subjective meanings within National Socialist ideology. Within what Uriel Tal called the *political faith* of the Nazi *Weltanschauung*, there was, in the words of Helmuth Krausnick, "an absolutization of the biological factor in all spheres of life."[29] On a mundane level, Wiepking-Juergensmann attributed the German feeling for landscape and relatedness to plants to "biological laws innate in our being."[30] At an ideological level, biology justified race: race was an inescapable law of nature,

---

[24] Reichel, *Der schöne Schein*, p. 120; Translation from German by the author.

[25] Reichel, *Der schöne Schein*, pp. 210-217.

[26] Uriel Tal, *Religion, Politics and Ideology in the Third Reich: Selected Essays* (Abingdon/New York, NJ: Routledge, 2004), p. 108.

[27] Margrit Bensch, *Die 'Blut und Boden' Ideologie: Ein dritter Weg der Moderne* (Berlin 1995), 39.

[28] Jonathan Petropoulos, *Art as Politics in the Third Reich* (Chapel Hill, NC/London: University of Carolina Press: 1996), p. 186.

[29] Cited in Tal, *Religion*, p. 22.

[30] Wolschke-Bulmahn/Groening, "The National Socialist Garden", p. 84.

its essence conditioned both by external nature (landscape) and internal nature (blood). Ernst Krieck argued that "race is [...] Nature's foundation for community, *Volk* and history."[31] Darre stated that "Race, as it rises above nature, always remains nature. [...] That which enables the higher development of a person is his *nature*."[32]

Biology also became the ground of spirituality, expressed by Jakob Wilhelm Hauer in his description of the German faith movement, which he led, as "an eruption from the biological and spiritual depths of the nation".[33] Gottfried Benn summed this up in a speech in 1932: "One of the classic perceptions of the post-Nietzschean epoch derives from Thomas Mann and reads: *"everything transcendent is animal, everything animal transcends."*[34] The spiritualisation of the biological could lead as far as the pantheism of Mathilde Ludendorff, whose book, *Erlösung von Jesu Christo* (Redemption from Jesus Christ)(1932), "sought to replace Jesus with a pantheistic adoration of nature: 'Because the entire world is permeated with God's grace, the German plants and animals are not soulless, like the servants of Yahweh'."[35]

Far more widespread was the substitution of "the metaphysical notion of biocentrism for the science-based one of logocentrism," adopted by Ludwig Klages, Jakob Wilhelm Hauer and other German Faith thinkers.[36] Central to metaphysical biocentrism is the importance of elementary symbols: "to Klages [...] each symbol *is* what it means; it *is* its meaning. It is metonymic and thus has psychic power." The faith of 'peoples of nature' [*Naturvölker*] was not word-based, but "image-based. Image-faith is knowable not through concepts, but only through symbols." A people's inner nature "is unlocked through images of gods, cults, symbols, mysteries and myths."[37] This analysis of the ancestors of European *Bildungsvölker* was by analogy also valid for National Socialist Germany, due to the widespread conviction, in Hauer's words, that God is "the power that resides

---

[31] Ernst Krieck, *Nationalsozialistische Erziehung*, (Osterwieck/Harz/Berlin: A.W. Zichfeldt Verlag, 1940), p. 5; Translation from German by the author.
[32] Bensch, *'Blut und Boden' Ideologie*, p. 46; Translation from German by the author.
[33] Cited in Steven E. Aschheim, *The Nietzsche Legacy in Germany 1890-1990* (Berkeley, CA /Los Angeles, CA /Oxford: University of California Press, 1992), p. 227.
[34] Cited in Ashheim, *Nietzsche Legacy*, p. 69.
[35] Richard Steigmann-Gall, *The Holy Reich: Nazi Conceptions of Christianity, 1919-1945* (Cambridge, MA: Cambridge University Press, 2003), 89.
[36] Karla Poewe, *New Religions and the Nazis* (New York, NJ/London: Routledge, 2006), p. 84.
[37] Poewe, *New Religions and the Nazis*, p. 85, pp. 87-8.

within the heart. [...] the heart is where the creative god unites with the upright and sacrificial person. [...] Within us presses the power of new emotion [*Ergriffenheit*], creative life from the holy depth of our *Volk*, from which all great things emerged on German soil."[38]

The rejection of logocentric rational thought and its replacement by symbolic metaphysical thought was crucial to many National Socialist ideologues and has a direct bearing on their perception of nature, which they regarded as inaccessible to rationality. Ernst Krieck argued that "rationalism cuts man from living [...] contact with cosmic reality, with nature, in other words from the womb from which Germanic man emerged and in which his great vitality was formed - in other words his essence (*sein Wesen*)."[39] In a 1940 *Gutachten* on the position of farmers in the Third Reich, Alfred Bauemler stressed as goals: "[...] to create anew in the light of consciousness a form of existence that hitherto resided in the unconscious [...] to nurture the irrational with rational means [...] proceeding from the purest impulses of the race."[40] Irrationality and intuition were championed as the means to access the powerful unconscious instincts of the Aryan race (while simultaneously many National Socialist political decisions were based on "supra-rational goals or motivations.")[41]

For Bauemler, as for Hitler and other National Socialist thinkers, myth was a key both to nurturing irrationality and to fostering action: "all people who have changed the face of the earth have lived from the depths of a myth."[42] Because "every true myth is a myth of the blood,"[43] each community of race has its own unique myth. Max Wundt, like other theorists of National Socialism, propagated its irrational blood myth, arguing that the nation could rid itself of the poison of Jewish rationalism and renew its vitality by returning to "the unconscious powers of nature which embrace us in the soil of our fatherland [...] speaking to us in the voice of our blood."[44] The SS *Glaubensbekenntnis* (Creed) stated that the return to roots would create "an eternal unity, a unity of blood and soil, soil and

---

[38] Poewe, *New Religions and the Nazis*, p. 89.

[39] From Ernst Krieck's address on 'Die Intellektuellen und das Dritte Reich', apparently delivered in 1938: in Tal, *Religion*, p. 8.

[40] Cited in Tal, *Religion*, p. 90.

[41] Tal, *Religion*, p. 88.

[42] Alfred Bauemler, *Alfred Rosenberg und der Mythus des 20. Jahrhunderts* (Munich: Hoheneichen Verlag, 1943), p. 68; Translation from German by the author.

[43] Bauemler, *Alfred Rosenberg und der Mythus des 20. Jahrhunderts*, p. 70; Translation from German by the author.

[44] Tal, *Religion*, p. 21.

nation, nation and race, race and God that is our Reich [...]". As Uriel Tal noted, "this mystique of union, as Himmler put it, was actually anchored in real nature and not metaphysics."[45] Yet real nature, as Carl Jung perceptively observed, "is not only aristocratic; it is esoteric."[46] Since myth itself embodies, as Ernst Cassirer noted in 1945, "demonic power,"[47] the Nazi myth of blood and soil concealed especially demonic consequences.

Friedrich Nietzsche had championed myth: "Without myth, every culture loses its healthy, creative natural vitality." He explicitly affirmed a demonic essence in myth: "the images of the myth have to be the unnoticed omnipresent demonic guardians." I believe that Nietzsche's myth-making was a vital element in the National Socialist understanding of nature – and far beyond. He was a crucial influence on Baeumler and, according to Steven Aschheim, "Nietzsche permeated every aspect of the *Glaubensbewegung's* [German faith movement's] counter-revolution." Aschheim points to "a profound affinity...the complicity of Nietzschean impulses within Nazism."[48]

In Nietzsche's view, modern "mythless man" had suffered "the loss of the mythical home, the mythical maternal womb[49]. In *Die Geburt der Tragödie aus dem Geiste der Musik*, published in 1872 while still intoxicated by Richard Wagner's music and vision, Nietzsche created the myth of Dionysus as the new, aristocratic spirit of future Germany. At its heart lies the ecstasy "arising from the innermost depth of man, indeed of Nature, on the annihilation of the principle of individuation." In that dionysian, annihilatory ecstasy, man is reunited with man and nature itself is reconciled with man in full beauty, as all boundaries are dissolved, "the gulfs between man and man give way to an overpowering feeling of unity, which leads back to the heart of nature."[50] Jung rightly considered that in Nietzsche's Dionysus, "the Deity appears in the garb of Nature."[51] For

---

[45] Tal, *Religion*, p. 33.
[46] Jolande Jacobi, ed., *Psychological Reflections: an Anthology from the Writings of C.G.Jung,* Bollingen Seies XXXI (New York: Pantheon Books, 1953), p. 272.
[47] Cited in Robert Jan van Pelt, "Bearers of Culture, Harbingers of Destruction: The Mythos of the Germans in the East", in Etlin, *Art*, 98-135, p. 120.
[48] Aschheim, *The Nietzsche Legacy*, p. 227, p. 320.
[49] Friedrich Nietzsche, *Die Geburt der Tragödie aus dem Geiste der Musik* (Augsburg: Goldmann Verlag, 1999), pp. 148-149.
[50] Nietzsche, *Die Geburt der Tragödie aus dem Geiste der Musik*, p. 28, p. 56; Translation from German by the author.
[51] Carl Gustav Jung, *Four Archetypes*, trans. R.F.C.Hull (London/Henley: Routledge & Kegan Paul, 1972), p. 52.

Nietzsche, a "return to Nature" is "an ascent – up into the high, free, even terrible nature and naturalness, such as plays with great tasks, is allowed to play with them."[52]

Few philosophers can have used imagery drawn from nature as extensively as Nietzsche, especially in his dithyrambic *Also sprach Zarathustra*. Zarathustra urges: "Create gardens!". References to the seasons, weather, sun, storm, mountains, certain animals abound. Yet they are repetitive, acting as symbols, signals of an inner mood or as eternally recurrent *leitmotifs*, a theatrical backdrop for the Dionysian myth which embodied "the oneness as genius of the species, indeed of Nature. Now the essence of Nature must express itself symbolically; a new world of symbols is necessary".[53] They never refer to the specific, the particular – which Nietzsche despised. He defined "the faith that only the particular is reprehensible, that in the whole everything redeems and affirms itself [...]" as "the highest of all possible faiths: I have baptized it with the name of Dionysus".[54] Thus, curiously, objective nature itself, in its infinite variety, specificity and beauty, disappears under Nietzsche's relentlessly projected inner vision, which emphasises above all the cruelty and overriding power of sexuality inherent within nature.[55]

One of Nietzsche's notebooks from the time of *Zarathustra* indirectly refers to this process: "To create the Superman after we have thought, indeed rendered thinkable, the whole of nature in terms of man himself" and then "to *break all your images of man* with the image of the Superman – this is Zarathustra's will."[56] Nietzsche's rational analysis is predicated on an essentially irrational, sustained mystical vision which anthropomorphises nature and, as it were, naturalises man as "refined beasts of prey [...]. Everything is an extension of their animality."[57] This eradication of any division between man and nature is central to Nietzsche's thought: "In reality, however, there is no such separation: [...] Man, in his highest and most noble capacities, is wholly nature and embodies its uncanny dual character. Those of his abilities which are awesome and considered

[52] Friedrich Nietzsche, *Götzen-Dämmerung*, in *Nietzsche Werke. Nachgelassene Schriften*, eds. Giorgio Colli & Mazzino Montinari (Berlin: Walter de Gruyter & Co., 1969), VI, vol.3, p.144; Translation from German by the author.

[53] Nietzsche, *Geburt*, p. 33; Translation from German by the author.

[54] Nietzsche, *Götzen-Dämmerung*, p.146; Translation from German by the author.

[55] Nietzsche, *Geburt*, pp. 56-58, p. 122.

[56] Erich Heller, *The Disinherited Mind* (London: Penguin, 1957; 1st. ed.1952), p. 142.

[57] Nietzsche, *Unzeitgemässe Betrachtungen*, III (Stuttgart: Alfred Kröner Verlag, 1955), p. 245; Translation from German by the author.

inhuman are perhaps the fertile soil out of which alone all humanity [...] can grow."[58] Nietzsche placed man "back among the animals. We consider him the strongest animal [...]. On the other hand, [...] Man is by no means the crown of creation: every living being stands beside him on the same level of perfection [...] man is the most bungled of all the animals, the sickliest, and not one has strayed more dangerously away from its instincts."[59]

It is central, too, to Nietzsche's opposition to Jews and to Christianity. The Jewish determination to survive, "to be *at any price"*, had led to their "radical *falsification* of all nature, all naturalness, all reality, of the whole inner world as well as the outer [...] out of themselves they created a counter-concept to *natural* conditions: they turned religion, cult, morality, history, psychology, one after the other, into an *incurable contradiction to their natural values*."[60] Christianity had invented an 'anti-natural' castrated 'god of the good alone' and then his counterpoint, nature: "Once the concept of 'nature' had been invented as the opposite of 'God', 'natural' had to become a synonym of 'reprehensible': this whole world of fiction is rooted in *hatred* of the natural (of reality!)."[61] Nietzsche saw Richard Wagner's music, above all his *Tristan*, as the healing of this radical Judaeo-Christian break with nature. In the spirit of Wagner's music lies the essence of the Dionysian return to nature, enabling us "to understand a joy in the annihilation of the individual."[62]

Since Theodor Adorno, like other commentators, noted "the deep interconnection between Richard Wagner and German supra-nationalism at its most destructive" and discovered "many elements of rubber-stamped Nazi doctrine in Wagner's theoretical writings,"[63] it is worth examining Wagner's views on nature in some detail. In a letter of December 1851 he described his own politics as "nothing more than the bloodiest hatred of our entire civilisation, contempt for everything that stems from it, and yearning for Nature [...]. Only the most terrifying and most destructive revolution can make out of our civilised beasts 'humans'

---

[58] Cited in Walter Kaufmann, *Nietzsche: Philosopher, Psychologist, Antichrist* (Princeton, NJ: Princeton University Press, 1974; 1st. ed. 1950), p. 178.

[59] Friedrich Nietzsche, *Der Antichrist* (Augsburg: Goldmann Verlag, 1999), pp. 20-21; Translation from German by the author.

[60] Nietzsche, *Antichrist*, p. 31; Translation from German by the author.

[61] Nietzsche, *Antichrist*, pp. 22-23.

[62] Nietzsche, *Geburt*, pp. 109-110; Translation from German by the author.

[63] Theodor W. Adorno, *Essays in Music,* selected by Richard Leppert, trans. by Susan H. Gillespie (Berkeley, CA/Los Angeles, CA/London: Univ. of California Press, 2002), pp. 374-5.

again."[64] He, too, had argued that within Nature had lain 'an ever-new creative force', accessible through the *Volk*'s roots to the 'soil of Nature' through instinctiveness.[65] In pre-Christian Greek religion, the *Volk*, marvelling at Nature's workings, had condensed "the many-membered show of Nature into a God, and finally its God into a Hero. In this hero it learns to know itself."[66] Yet "the Judaeo-Christian [miracle] tore the connexion of natural phenomena asunder, to allow the Divine Will to appear as standing *over* Nature"[67] The Judaeo-Christian "abstract God [...] *ruled* the world –God, who had *made* all Nature for the gory of his name. From that time forward, man's affairs are governed by the '*incomprehensible will*' of God; no longer by the instinct and necessity of Nature". Judaeo-Christian culture had not "sprung upwards from the nether soil of Nature," but had been "poured down [...] from above, from the Heaven of the priests."[68] Uprooted from the creative, unconscious instincts of nature, it was barren, barbaric, creating only "cripple-like monstrosity."[69]

Wagner believed himself to be uniquely able to reverse this historical disjunction with nature. Stylising himself as 'the Poet', he argued that the Poet, the '*knower of the unconscious*', might achieve "loving intercourse with Nature" and become "lord of Nature."[70] In Wagner's own projected '*Artwork of the future*', the entire [German] *Volk* would become "*one* – [...] knowers of the unconscious [...] *blissful men.*"[71] His *Artwork* (from the *Ring* cycle onwards) would effect "the conscious reunion of Nature with man"[72] by creating "the *new religion.*"[73]

---

[64] Cited in Hartmut Zelinsky, "Verfall, Vernichtung, Weltentrückung: Richard Wagners antisemitische Werk-Idee als Kunstreligion und Zivilisationskritik und ihre Verbreitung bis 1933" in Saul Friedländer and Jörn Rüsen, eds., *Richard Wagner im Dritten Reich*, (Munich: Verlag C.H.Beck, 2000), 309-341, p. 313; Translation from German by the author.

[65] Richard Wagner, "Oper und Drama (1851)", in Dieter Borchmeyer, ed., *Richard Wagner. Dichtungen und Schriften: Jubiläumsausgabe in 10 Bänden* (Munich: Insel Verlag, 1963), vol.7, p. 259.

[66] Wagner, "Oper und Drama (1851)", p. 61; Translation from German by the author.

[67] Wagner, "Oper und Drama (1851)", p. 207; Translation from German by the author.

[68] Richard Wagner, "Art and Climate (1850)", in *The Artwork of the Future and other Works*, trans. W. Ashton. Ellis (Lincoln, NE/London: University of Nebraska Press, 1993), p. 255.

[69] Wagner, "Art and Climate (1850)", p. 256.

[70] Wagner, "Oper und Drama", 259, p. 280; Translation from German by the author.

[71] Richard Wagner, "The Artwork of the Future (1849)", in *The Artwork of the Future and other Works*, trans. W. Ashton. Ellis (Lincoln, NE/London: University of Nebraska Press, 1993), p. 77.

[72] Wagner, "Art and Climate", p. 253.

Redemptive elements of Wagner's essentially pagan 'new religion' included: the recognition that "the *only* God indwells in us and in our unity with Nature" and therefore "we are God himself: for God is the knowledge of self"[74] and encouraging "the still unbroken nature-force of Race [...] *pride* [...] of race" and arousing anti-semitism, "the re-awakening of an instinct lost to the German nation. [...] a German instinct."[75] Since the Jew represents "this [...] freak of Nature", the "instinct" of Jew-hatred – "our natural repugnance against the Jewish nature" – is natural and demands a "war of liberation."[76] Because "unconscious agency is the agency of nature," intellect – "obsolete intelligence" – must be "completely annihilated by and in itself." For only when all know "that we must yield ourselves, not to our intelligence, but to the necessity of nature," only when all are "brave enough to deny [...] intellect, shall we obtain from natural unconsciousness [...] the force to produce the new."[77]

The eventual demise of the Jew, whose fate was envisaged as 'self-annulment' and '*Untergang*' (destruction),[78] would result from the demise of the Jewish God. Wagner wrote in 1881: "Only when [the Jewish God] [...] no more can find a where or when to lurk among us, will there also be no longer – any Jews."[79] Shortly after Wagner's death in 1885, Joseph Popper-Lynkeus perceptively noted the murderous potential within the ideology of Wagner's antisemitic disciples: "Those, who melt with sheer love for animals would in all tranquillity allow the murder of Jews."[80] Significantly, in Wagner's *Siegfried*, the Aryan God-hero Siegfried can understand the voice of Nature (Woodbird) only when he has experienced "*des Blutes Genuß*" (the enjoyment of blood), tasting the blood of Fafner, whom he has just murdered. The Woodbird then reveals to him his destiny to become ruler of the world (Act II, scene 2) and then in the following

---

[73] Wagner, *Artwork*, p. 155.

[74] Richard Wagner, "Jesus of Nazareth (written 1848-9, 1st.pubd.1887)", in *Jesus of Nazareth and other Writings*, trans. W. Ashton. Ellis (Lincoln, NE/London: University of Nebraska Press, 1995), pp. 311-2.

[75] Richard Wagner, "Know Thyself (1881)", in *Religion and Art*, trans. W. Ashton. Ellis (University Lincoln, NE/London: University of Nebraska Press, 1994), p. 269, p. 272.

[76] Richard Wagner, "Judaism in Music (1850)", in *Judaism in Music and other Writings*, trans. W. Ashton. Ellis (Lincoln, NE/London: University of Nebraska Press, 1995), pp. 81-3.

[77] Wagner, "Sketches and Fragments (1849-51)", in *Artwork*, p. 345.

[78] Wagner, "Judaism in Music", p. 100.

[79] Wagner, "Know Thyself in Religion and Art", p. 274.

[80] Cited Zelinsky, "Verfall", p. 326; Translation from German by the author.

scene explicitly approves of his murder of Mime, his archetypal representative Jew.[81]

Nietzsche famously proclaimed his own 'murder' of God in *The Gay Science* – in that fateful year, 1881. Key for Nietzsche was the notion that "the murderer of God must himself become God – that is, an *Übermensch* – otherwise he will sink into banality."[82]

National Socialist ideologues reflected and intensified this mythic vision of (German) man and *Volk*, reunited with nature's primal force through instinct and through dionysiac replacement of the abstract Jewish God: transcendance replaced by biology. If biology explained the basis for German creativity and re-awakened spirituality, Nazism also found, as Josh Cohen recently stated, "in biology the basis for the determination, isolation and eventual annihilation of bare life".[83]

Jews were perceived as having an unchanging evil nature. In Baeumler's words: "one must take the opponent seriously, one must recognise that these people *must* behave thus, since they *are* thus, and that they will therefore never change themselves."[84] In 1903 Otto Weininger condemned Jews as being without awe before Nature, without mystery or true mysticism: "Of the divine *within* man, of the 'God, who dwells within my breast', the true Jew knows nothing."[85] Worse, according to National Socialist ideology, Jews destructed the true Germanic nature, intented on uprooting Germans "from the sources of our strength,"[86] and on destroying the German 'personality': "therefore the German must become the world-historical opponent of the Jew."[87]

Robert Pois wrote of Hitler and the core National Socialist elite: "In a very real sense [they] perceived themselves as being virtual mediators between man – or at

---

[81] Zelinsky calls Mime one of Wagner's 'Judenkonzeptionen': ibid., p. 318.

[82] Rüdiger Safranski, *Nietzsche: A Philosophical Biography*, (London: Granta Books, 2003), p. 271. (1st.pubd. as *Nietzsche: Biographie seines Denkens* (Munich/Vienna: Carl Hauser Verlag, 2000).

[83] Josh Cohen, *Interrupting Auschwitz: Art, Religion, Philosophy* (New York, NJ/London: Continuum, 2003), p. 8.

[84] Baeumler, *Alfred Rosenberg*, p. 20.

[85] Otto Weininger, *Geschlecht und Charakter*, (Vienna/London: Wilhelm Braumüller, 1925; 1st.ed.1903), pp. 416-7, p. 420; Translation from German by the author.

[86] Theodor Fritsch, *Handbuch der Judenfrage: Die wichtigsten Tatsachen zur Beurteilung des Jüdischen Volkes* (Leipzig: Hammer Verlag, 1939), p. 103; Translation from German by the author.

[87] Baeumler, *Alfred Rosenberg*, p. 47.

least a group of men – and a savage natural world."[88] Only they embodied, in Gregor Strasser's words, the "one 'correct' spirit [...] the spirit which through man, in God's image, animates eternal nature! [...] this spirit is in us, in the idea of National Socialism. And [...] it and no other will build the *Millenial Reich*!" That spirit was the awakening of the instinct of antisemitism and, in party member Wilhelm Scherer's words: the "struggle [...] against the Jewish spirit within the German Volk's soul."[89]

Hitler was described by Jung in October 1938 as "a kind of medicine-man or shaman [...]. He reflected the unconscious of the Germans."[90] In pre-modern pagan religions the only one to mediate between the gods and the earth and to appear to control nature was the shaman, able to penetrate the three spheres of sky, earth and the lower realms. The shaman was healer, miracle-worker, psychopomp and the sole 'great master of ecstasy', specialising in trance states in which, crucially, he is able to communicate with the dead 'demons' and 'nature spirits'. The shaman "not only directs the community's religious life but, as it were, guards its 'soul'."[91]

Leni Riefenstahl's film *Triumph of the Will* stylised Hitler's plane as swooping down from the heavens to the earth; at key moments in the Nuremberg Party rallies Hitler appeared to ascend from the depths to a raised dais, while at others he marched alone across a vast distance, the attention of hundreds of thousands focussed exclusively upon him. Bauemler stressed that Hitler, "the soul-awakener of the nation", had a "virtually magical effect" on all who came in contact with him.[92] Himmler thus described Hitler spoke to a convention of SS leaders in 1935: "[He]has raised this Germany once again [...]. He [...] wants to return us to the source of the blood, to root us again in the soil – he seeks again for strength from sources which have been buried for 2000 years."[93]

In *Mein Kampf*, Hitler affirmed that "the *volkish* philosophy of life corresponds to the innermost will of Nature." He argued that "The Jew is the anti-man, the

[88] Pois, *National Socialism*, p. 40.

[89] Cited in David Redles, *Hitler's Millennial Reich: Apocalyptic Belief and the Search for Salvation* (New York, NJ/London: New York University Press, 2005), p. 75, p. 64.

[90] Gerhard Wehr, *Carl Gustav Jung: Leben, Werk, Wirkung* (Munich: Kösel-Verlag, 1985), p. 289; Translation from German by the author.

[91] Mircea Eliade, *Shamanism: Archaic Techniques of Ecstasy*, trans. Willard. R. Trask, (London: Arkana Penguin, 1989; 1st.ed. 1964), pp. 4-8.

[92] Baeumler, *Alfred Rosenberg*, p. 23.

[93] Baeumler, *Alfred Rosenberg*, p. 23.

creature of another god [...]. He is much further from the beasts than we Aryans. He is a creature outside nature and alien to nature." In the 1930s his metaphor for opposing Jews was that of a gardener eradicating weeds to keep a garden healthy.[94] In *Mein Kampf*, Hitler had noted that "men [...] owe their higher existence [...] to the knowledge and ruthless application of Nature's stern and rigid laws."[95] The Holocaust and Nature are, indeed, intimately connected.

[94] Cited in Redles, *Hitler's Millennial Reich*, p. 71, p. 67, pp. 62-3.
[95] Cited in Pois, *National Socialism*, p. 38.

In Response to Margaret Brearley

Rochelle L. Millen

The embeddedness of – and ambiguity towards – Nature in Nazi ideology is meticulously articulated in Margaret Brearley's essay. The notion that Nature is the source of human power and creativity and must be preserved in its pristine state was counterbalanced, in Nazi principles, by the often radical exploitation of natural resources for military or even artistic purposes. Similarly, the Nazi vision of the perfect Aryan body trained to fight for – or biologically reproduce – that ideal was offset by the willing risk of mass death in submission to that very ideal.

What I wish to focus upon here is the concept of Nature itself, especially as Nietzsche understands it. For Nietzsche admires – indeed affirms – vigor, vitality, and the animation of inner strength as intrinsic to values deriving from Nature. He celebrates an enhanced feeling of living, of life without certainty, of existing on the edge in a state of exhilaration. Security is a betrayal of whatever meaning is in human existence. While admiring the Greeks, Nietzsche criticizes Socrates[1], but likes the pre-Socratics; Heraclitus's notion of the world as constantly shifting its center fits well in Nietzsche's perspective. As beauty in a work of art, living is its own end; as in the conflicts of natural forces, the self flourishes when it struggles. Human physicality is a gift to be cultivated as one would nurture a rare rose.

Thus it is, according to Nietzsche, that when spirituality becomes more important than vitality, Nature has been corrupted. The Socratic notion of the soul imprisoned in the body, longing for its liberation, creates a false hierarchy of priorities. Nietzsche's paradigm is the full integration of body and soul, of vigor and intellect.

This model posits a radical individuality, an abrogation of the cultural consequences of Hebrew Scriptures, Greek philosophy, and the Pauline amalgamation of them both, *i.e.* Christianity. The moral implications of breaking the covenant, the priority of reason, and the concept of original sin, Nietzsche sees as inver-

---

[1] See Weaver Santaniello, "Socrates as the Ugliest murderer of God," in Weaver Santaniello, ed., *Nietzsche and the Gods* (New York, NJ: SUNY Press, 2001), 73-86.

sions of natural values. For each of these, especially Hebrew Scriptures[2] and Pauline Christianity, demands obedience to a transcendent, objective Deity. Submission or obedience, however, connoted the absence of struggle, to Nietzsche, the quiescence of passivity, the lack of insight into the self. Christianity's negative view of the body as the source of sin,[3] its abnegation of physical pleasure, was *anathema* to Nietzsche.

Nietzsche detested what he saw as the heteronomous moral structure deeply entrenched in European Christian culture, viewing human nature instead as encapsulated in a radical autonomy. That is why he despised the modern state and the rampant nationalism of the nineteenth century: parameters limited the individual, mandating obedience rather than struggle, passivity instead of life-affirmation. In the words of Tim Murphy: "it is clear that much of what Nietzsche was trying to say in his treatment of Jews and Judaism was as much – if not more – about the German-Christian construction of modern European identity than it was about the Jews themselves."[4]

At this point, two difficulties emerge. First, Nietzsche's philosophy fails to account for human community. In its emphasis on radical individuality and struggle, the need for a framework of relatedness is not explored. Even within its resistance to systemization, this lacuna is significant. Can there be human connection free from hierarchy and authority? Perhaps not in any family structure or political entity Nietzsche could envision. And second, the Nazi state which elevated Nietzsche to its "intellectual *Führer*"[5] and "prophet of a new legal order"[6] clearly represented that which Nietzsche abhorred: group conformity, submission to authority, denigration of individuality, and overarching rationalism. Yet Nazism, as Brearley notes, continues to be regarded as an ideology, however selectively mediated, carved from Nietzschean constructs. The nature of Nature in Nazism becomes a corrupted version of Nietzsche's vision of the vital individ-

---

[2] It is highly unlikely that Nietzsche had any knowledge of *midrash* or rabbinic texts, both of which convey some of the subtlety and ambiguity regarding human nature Nietzsche valued.

[3] See Elaine Pagels, *Adam, Eve, and the Serpent* (New York, NJ: Vintage Books, 1989), especially Chapter VI, "The Nature of Nature."

[4] Tim Murphy, "Nietzsche's Narrative of the 'Retroactive Confiscations' of Judaism," in Weaver Santaniello, *Nietzsche and the* Gods (Albany, NJ: State University of New York Press), 3-20.

[5] Quoted by Steven E. Aschheim, *The Nietzsche Legacy in Germany, 1890-1990* (Berkley, CA: University of California Press, 1992), p. 241.

[6] Aschheim, p. 242.

ual. The lack of a philosophy of the multitude, of the nature of community, of the relationship of society to Nature, were key factors resulting in a mass political movement which unleashed a destructiveness contrary to Nietzsche's original goals.

One might argue, with Yirmiyahu Yovel,[7] that Nietzsche's positions on Nature, Christianity, and Judaism were left open to abuse, "for the mental revolution he sought did not take place, while his ideas were generalized [and] vulgarized." With certain ideas, Nietzsche "was playing a dangerous game [...]. The irony of speaking ironically to the vulgar is that the speaker himself may end up the victim of an ironic reversal [...]. Nietzsche as a master of the art should have anticipated the ironic fate of ironizers."[8] Yovel argues that Nietzsche's psychological model of the human being was for the few (the Platonic philosopher-kings?) and not for the herd; thus the abuse of his visions, when incorporated into a nationalist political philosophy, founded on race and an untrammeled sense of Nature.

---

[7] Yirmiyahu Yovel, "Nietzsche and the Jews: The Structure of an Ambivalence," in Jacob Golombh, ed., *Nietzsche & Jewish Culture* (New York, NJ: Routledge, 1997), 117-136.
[8] Yovel, pp. 126-127.

In Response to Margaret Brearley

Peter Haas

This article brings out in an exceptionally clear way the Nazi's dichotomous view of, and relationship with nature. Brearley clearly demonstrates that the Nazis saw themselves both as part of nature, drawing their very power and life from it, and at the same time as having a duty to transcend and overcome the savage and beastly. It is thus no surprise that Nazi policies toward nature seemed to have unresolved tensions, as the opening paragraphs of this chapter illustrate. This mixture of attitudes, as the chapter also makes abundantly clear, was embedded deeply in the Nazi mythic structure of reality.

It is this mythic definition of nature to which I want to turn my attention, because the very logic of myth suggests a way of moving further into an attempt to explain the Nazi phenomenon. In the study of religion, 'myth' is the foundation narrative of a community, and as such stands at the very center of that community's being. A society's story about itself doesn't only define its origins, but maybe more importantly, also its boundaries. A myth, in short, explains why that community is what it is, and often why others are different and even inferior. It is through such mythic retellings of origin and history, that a community relates its understanding of itself to its own members, and that it defines for its members how the 'Other' is to be understood and treated. In so far as a myth is taken to reflect the really real out there, it is in fact that what makes a group of individuals into a society or a community in the first place. As such, the myth also defines what is right and what is wrong, or in Durkheim's words, what is 'sacred' and what is 'profane'. Myths, then, are powerful shapers of self-identities and ethics. Myths provide its audience with the words and symbols for getting at the True and the Good.

As a 'sacred' text, however, a myth like any other sacred text needs interpretation. No text, whether oral or written, contains within itself its own authoritative and unambiguous interpretation. To the contrary, it is often its very ambiguity and multivaliency, which gives a myth the flexibility to sustain itself over the vicissitudes of time. This is why any community eventually produces a cadre of authoritative interpreters. Without interpretation into the terms of everyday life, a myth becomes otiose, of no real relevance. And without interpretation into new forms as times change, a myth becomes obsolete. By casting at least part of the

Nazi program in terms of its myths, Dr. Brearley gives us anther route into the interpretation of the Nazi state.

By establishing nature as part of the central narrative structure of their myth, the Nazis by the sheer logic of nature, moved scientists into the central interpretive role. Scientists, after all, are by profession master interrogators of nature. For the nineteenth and twentieth centuries, the scientist in his lab, or out in the field, became the paradigmatic figure of how knowledge is to be gained. It thus is quite natural that by the second half of the twentieth century in North America, scientists were fully engaged in casting religion as little more than superstition, and declaring that there was a war between religion and science and that science had to win out. The result was that by the time of the rise of the Nazis, it was taken for granted by large swaths of the population that it was science that held the keys to knowledge and thus to a sort of secular salvation, and that religion, with its fantastic stories, obscure rituals, and wordy preachers, had a clearly secondary status.

The irony of course, is that the rise of this 'Scientism' really was the displacement of one series of myths with another, one priesthood of knowledge with another, and one set of self-evident ethics with another. It is undoubtedly the rise of the 'cult' of science that accounts for the significant numbers of physicians and other scientists who became implicated in the horrors of the Shoa.

If the scientists were the source of new knowledge, constituting a sort of new 'Magisterium', then it was the state bureaucracy that was the primary Acolytes, who turned the new revelations extracted from nature into concrete policies and liturgies. It was their task as well to explain the world in terms of the taxonomy the myth provided, and to account for events in ways that worked within the boundaries of the myth – that is in a way that did not threaten the inner coherence, outer boundaries or comprehensiveness of the mythic reality Nazis scientists and mythographers were propounding. The bureaucrats followed, of course, their own rigid logic in the enactment of laws drawn up in light of the categories proposed by the myth. Looking in from the outside, for example, we are struck by the intricate mechanisms set in place to determine racial categories and to control their social lives accordingly. We are astounded by the painstaking exactitude exercised in the racial courts, and the precision with which percentages were assigned to various racial admixtures in any one individual. Following Brearley's lead, however, we can see that it is in precisely these details that the myth must either work, or be shown to be incoherent. A myth based on nature, must explain nature in all its messy varieties, or suffer from internal inconsistency and even cognitive dissonance.

The idea of reading the Nazi phenomenon in terms of its basis on a certain type of myth, explains much of the character of the Nazi state. While on the one hand a myth has to be multivalent and flexible to enable its community to survive, it also has to be reduced to a relatively uniform series of rituals or other practices, so that the community can periodically reinforce their subscription to the myth. Regular reinforcement, in this case propaganda, is necessary. And while, on the other hand, any myth must allow for a certain amount of diversity and non-uniformity in its implementation, at some point divergence from a generally accepted norm becomes unacceptable and even dangerous, to the point to where it has to be exorcised. It is for this reason, I submit, that the Nazi state had so many characteristics that recall religious activities: marches, holy days, martyrs, sacred precincts and all the rest. A kind of religion of nature presided over by scientists came into being in the Nazi state.

In the light of these considerations, I think it becomes clear why the stakes in the Nazi myth became so high. If the ultimate source of the definitive myth is the very nature of Nature itself, then the importance of the sacred community rises to world – even cosmic – proportions. In some sense we can look at the myth of nature as a version of the 'grand myth' of an apocalyptic battle between the forces of light, order and civilization (the sacred), and those of darkness, chaos and so the demonic. Not all communities have mythic underpinnings of this scope, but for those that do, the importance of a positive outcome is so great that almost any means are justifiable.

What Brearley has given us, of course, is a highly intellectualized version of the Nazi myth, as drawn from figures like Wagner and Nietszche. But the myth in its more general form came to permeate the instruments of the Nazi state. By promulgating a myth that encompasses the fullness of Nature, the Nazis were able to establish a battle of ultimate significance. The mythic analysis presented in this chapter helps to clarify the source from which this power was drawn.

In Reply to My Respondents

Margaret Brearley

In his perceptive response, Peter Haas rightly stresses the importance of (non-Jewish) scientists as central interpreters of Nazi myth. In 1935 Philipp Lenard advocated 'Aryan physics'; in 1936 Bernhard Rust, *Reichserziehungsminister* (Minister for Education) argued that rejection of Jewish 'objective' science had 'transformed the inner life of the German people', helping to forge an "organic unity" between science and the Volk; and in the same year Ernst Krieck said that: 'it can be fully demonstrated...that...every worthwhile achievement in the sphere of the natural sciences, no less than in the sciences of culture, has been intimately bound up with the...racial characteristics of the people concerned'. [1]
Medicine and biology were arguably the sciences most permeated by Nazi political thought, which is one reason why 'doctors were more enthusiastic Nazis than members of any other profession'. [2] Robert Jay Lifton described the Nazi state as a "biocracy"; the Darwinian term 'selection' was widely used, since 'the Nazis sought to take over the functions of Nature...and God...in orchestrating their own "selections", their own version of human evolution'.[3] Rudolf Hess claimed in 1934 that 'National Socialism is nothing but applied biology'.[4] But it was surely less 'undoubtedly the rise of the 'cult' of science that accounts for the significant numbers of physicians and other scientists implicated in the horrors of the Shoa' (Haas) than the cult of 'scientific racism' – a form of pseudo-science which had underpinned German practice of genocide long before the Holocaust.
Already German colonial policy in German South West Africa had used principles of eugenics and racial hygiene to justify the virtual eradication of the Christian Herero and Nama tribes, whose cattle and extensive lands German settlers wished to expropriate. Captain Maximilian Bayer, who had volunteered for service against the Herero in 1904, used pseudo-scientific 'laws of nature' to validate mass slaughter:

---

[1] Peter Watson, *The German Genius: Europe's Third Renaissance, the Second Scientific Revolution and the Twentieth Century* (London: Simon & Schuster UK, 2010), p. 652-3.
[2] Ibid., p. 661.
[3] Robert Jay Lifton, *The Nazi Doctors: Medical Killing and the Psychology of Genocide* (New York; Basic Books, 2000: first published 1986), 17.
[4] Ibid., p. 31.

> Out Lord has made the laws of nature so that only the strong have a right to continue to exist in the world, and so that the weak and purposeless will perish in favour of the strong. This process is played out in a variety of ways, like, for example, the end of the American Indians, because they were without purpose in the continued development of a world that is striving towards a higher level of civilisation; in the same way the day will come when the Hottentot [Nama] will perish, [it will] not [be] any loss for humanity because they are after all only born thieves and robbers, nothing more.[5]

Fascinatingly, Raymond E.Fancher has argued that Francis Galton, the founder of eugenics, developed his theories partly under the influence of racial attitudes acquired during his stay in what later became German South West Africa in 1850-1. Following that visit, Galton expressed strongly racist contempt for the Nama ('children') and for the Hereros ('so filthy and disgusting in every way....a greedy, heartless set of savages').[6]

German scientists committed to *Rassenhygiene*, as the foremost German eugenicist, Alfred Ploetz (1860-1940), termed eugenics, were as early as 1891 co-opted to *Rassedienst* (service of the race) which, as Paul Weindling recently noted, 'opened the way for physicians to stigmatize not just otherness but also a range of medical conditions, behaviours and identities, as a pathological threat to the body politic. Biology and medicine were permeated by the language of the state as an organism'. [7] Moreover, *Rassenhygiene* was itself co-opted to serve forms of mythology later utilised by Nazism; within the Racial Hygiene Society, founded by Ploetz in 1905, Ploetz himself founded in 1912 'a secret inner core' dedicated to Nordic body culture (*Der Bogen*), which after World War I continued as the *Widar-Bund*, dedicated to Widar, the Nordic god of light,[8] There were German eugenicists who specifically rejected racism, as Marius Turda has

---

[5] Maximilian Bayer, *Der Krieg in Südwestafrika und seine Bedeutung für die Entwicklung der Kolonie* (Leipzig: Verlag von Friedrich Engelmann, 1906), p. 9; Cit. David Olusoga and Casper W. Erichsen, *The Kaiser's Holocaust: Germany's Forgotten Genocide and the Colonial Roots of Nazism* (London: Faber and Faber, 2010), 133.

[6] Raymond E. Fancher, "Francis Galton's African Ethnology and its Role in the Development of his Psychology", *British Journal of the History of Science* (1983), vol. 16, 67-79, p. 74-75.

[7] Paul Weindling, "German Eugenics and the Wider World: Beyond the Racial State", in Alison Bashford and Philippa Levine, eds., *The Oxford Handbook of the History of Eugenics* (Oxford; Oxford University Press, 2010), 315-331, p. 31 (316).

[8] Ibid., p. 319.

pointed out, but by the late 1930s the majority had committed themselves to a sinister version of the path more innocently suggested in 1936 by Julian S. Huxley, who foresaw that 'eugenics will inevitably become part of the religion of the future, or of whatever complex of sentiments may in the future take the place of organized religion'.[9]

Rosenberg was one of the most prominent Nazi ideologues who explicitly utilised *Rassenhygiene.* He argued, for example, that 'No feelings of or for Jewish 'humanity' must in the future be carried out, because of our Nordic racial and hygiene views'. [10] Haas notes that 'a kind of religion of nature...came into being in the Nazi state'. Rosenberg explicitly champions this religion of nature, which replaces the sacraments of Christianity and stands in antithesis to Judaism:

> Today a new faith is awakening – the Myth of the blood: the belief that to defend the blood is also to defend the divine nature of man in general. It is a belief, effulgent with the brightest knowledge, that Nordic blood represents that *Mysterium* which has overcome and replaced the older sacraments....The Jewish concept of God wills [a destruction of natural law]....The sun myth of the Aryan is not only transcendental but also a universal law of nature and biology. [11]

Rosenberg interprets innate characteristics of Nazism as directly reflecting Nature itself: Nazi dynamism represents 'the eternal flux of nature'[12], its focus on intuitively perceived will 'is the driving momentum in the whole of nature'.[13] Indeed, the 'Germanic essence' embodied within Nazism not only affirms 'the divine nature of man' but the divine essence of nature, its very unity with God: 'The "sacred union" of God and nature is the primal ground of our being'.[14]

While this might be reminiscent of references in the American Constitution of July 1776 to 'the Laws of Nature and of Nature's God', the god intended by Rosenberg is explicitly very different from 'the Jewish God'[15] or the vaguer

---

[9] Marius Turda, "Race, Science and Eugenics in the Twentieth Century", in Bashford/Levine, op.cit., 62-79 (67, 72).

[10] Alfred Rosenberg, *The Myth of The Twentieth Century: An Evaluation of the Spiritual-Intellectual Confrontations of our Age'*, third edition (1931), translated by Vivian Bird (Sussex: Historical Review Press, 2000), p. 377.

[11] Ibid., p. 65-66, 79, 82.

[12] Ibid., p. 78.

[13] Ibid., p. 208.

[14] Ibid., p. 145.

[15] Ibid., p. 74.

ethical deity of the Deists. Rosenberg is clear that, while Odin might be dead, 'the driving power behind all is – the will' and 'the German mystic discovered "the strong one from above" in his own soul'.[16] The implication is that, for the new Nazi German, communion with god and nature is through inner identification with Hitler's will.

In her finely drawn response relating to aspects of Nietzsche's thought, Rochelle Millen argued that the Nazi state 'clearly represented that which Nietzsche abhorred'. Yet only, perhaps, in part; like Nietzsche, the Nazi state championed irrationality rather than 'overarching rationalism', and ecstatic identification with Hitler's will is, perhaps, not so very different from Nietzsche's intoxicating *Rausch*. Rochelle Millen perceptively notes that Nietzsche is concerned with autonomous individual freedom and 'failed to account for human community'. This was, perhaps, inevitable, given the Dionysian role of Eros and euphoria both in Nietzsche's philosophy and, more ambiguously, in his life (homoerotic Eros on the one hand, altered consciousness through hashish and chloral hydrate on the other).

Yet one could argue that Nietzsche did indeed anticipate a human community based on his thought. His 'noble human being' is declared noble not for any good or noble actions, but for overriding *faith* in himself, his egoism aristocratic and untrammelled. He affirms a community of similarly 'noble' individuals: 'a good and healthy aristocracy...accepts with good conscience the sacrifice of untold human beings who, *for its sake*, must be reduced and lowered to incomplete human beings, to slaves, to instruments'.[17] As Alexander Nehamas noted, Nietzsche accepted 'that sublimated cruelty...may well erupt in its crudest, most horrifying forms...'[18]

> 'One may be quite justified in continuing to fear the blond beast at the core of all noble races and in being on one's guard against it: but who would not a hundred times sooner fear when one can also admire that *not* fear but be perma-

---

[16] Ibid., p. 145, 137.

[17] Friedrich Nietzsche, *Beyond Good and Evil*, trans. Walter Kaufmann (New York: Random House, 1974), par. 258.

[18] Alexander Nehamas, "Nietzsche and 'Hitler'", in Jacob Golomb and Robert S. Wistrich (eds.), *Nietzsche, Godfather of Fascism? On the Uses and Abuses of Philosophy* (Princeton, NJ; Princeton University Press: 2002), 90-106, p. 99.

> nently condemned to the repellent sight of the ill-constituted, dwarfed, atrophied, and poisoned?' [19]

Daniel Conway recently pointed out that, when writing *Beyond Good and Evil* in the late 1880s, Nietzsche 'had empire on his mind' [20]– not the fast burgeoning German empire in Africa, but a cultural empire within Europe. He called for 'the cultivation of a new caste that will rule Europe'. [21] As Conway noted, 'Nietzsche apparently believes that the goal of pan-European renewal would justify *any* means necessary, including, as he says, "slavery" and the deliberate " cultivation of tyrants" (*Beyond Good and Evil*, par. 242)'.[22]

Nietzsche's vision of the noble human being was indeed for the few, not the herd; but he seems to have anticipated with equanimity the future subjection of the herd, for which he had nothing but contempt, to the overriding will of the autonomous, fearsome, 'noble' few.

---

[19] Nietzsche, *On the Genealogy of Morals*, trans. And ed. Walter Kaufmann (New York: Random House, 1967), I: p.11.

[20] Daniel W. Conway, "*Ecce Caesar*: Nietzsche's Imperial Aspirations" in Golomb/Wistrich, op.cit. 173-195, p. 175.

[21] Nietzsche, *Beyond Good and Evil*, par. 251.

[22] Conway, op.cit. 179.

Land, Nature, and Judaism:

Post-Holocaust Reflections

Rochelle L. Millen

In 1916, Martin Buber founded the journal *Der Jude*, of which he was to remain editor until it ceased publication in 1928. Noting in his initial essay that Gabriel Riesser – the early advocate of German-Jewish emancipation – had in 1832 founded a periodical of the same title, Buber explained the difference in purpose between them. Riesser had intended his periodical for individual Jews struggling with the issue of equal civic status before the law. "We give our organ the same name," Buber stated, "but we are not concerned with the individual, but with the Jew as the bearer and beginning of nationhood."[1] In stating this, Buber articulated not only the political Zionism spearheaded by Theodore Herzl,[2] but also the religious/cultural Zionism deeply ingrained among Eastern European Jewry. Buber proclaimed the profound connection between the land of Israel and the religion of Judaism at a time when issues of civic emancipation of Jews in Germany seemed smoother and less problematic. After all, in 1916, 100, 000 Jews were fighting for Germany in WW I; [3]Jews were German patriots *par excellence.* Yet Buber's comment pointed to an anomaly: if Judaism is more than a religion – indeed it is a peoplehood as well – than how does German patriotism fit in? Can Jews be accepted as part of the German people and state when they are also part of the Jewish nation?[4]

---

[1] Arthur A. Cohen, ed., *The Jew: Essays from Martin Buber's Journal, Der Jude,* trans. Joachim Neugroschel (Philadelphia: Jewish Publication Society, 1980), p. 10.

[2] Buber met Herzl in Berlin in 1898, one year after the First Zionist Congress was held in Basel.

[3] Of these 100,000, 19,835 were from the greater Berlin area and 12,000 died on the front. See Raymond Wolff, "Zwischen Formaler Gleichberechtigung, Zionismus und Antisemitismus" in Manfred Hammer and Julis H. Schoeps, eds., *Juden in Berlin, 1671-1946: Ein Lesebuch* (Berlin: Nicolaische Verlagsbuchhandlung Beuermann GmbH, 1988), p.130. My thanks to Dr. Timothy Bennett for this source.

[4] Europe begins, in the early twentieth century, to deal with the question of whether a nation and a state are necessarily the same entity, leading to the development of treaties to protect minorities. See Robert Cover, *Narrative, Violence, and the Law: Essays by Robert Cover*, Martha Minow, Michael Ryan & Austin Sarat (eds.)(Ann Arbor, MI: University of Michigan Press, 1995), pp. 26-28. Also, Carole Fink, *Defending the Rights of Others: The*

Using a post-Holocaust perspective, this essay will explore the intertwining of Judaism with the land of Israel as a concrete expression of nature. Judaism differs from other religious traditions in the fact that its full realization, according to Hebrew Scripture and the rabbinic texts which interpret Scripture, occurs in a specific geographic location: the land of Israel.[5] Scripture articulates the founding of a people who share a common fate and destiny;[6] whose ethical, moral, and theological models, while universal in an abstract sense, are rooted in the people's concrete, lived, historical experiences. Nineteenth century German Romanticism, with its emphasis on German folk myths and German soil, similarly sprouts from an understanding of shared experiences. Indeed, nationalism as a general European movement in the nineteenth century, while understood by some as a common cultural design, led to stronger fragmentation and antagonisms than earlier.[7] The underlying notion of nationalistic movements was a shared common past among a particular group. At the same time, however, the rationalism of the eighteenth century stressed the common roots of humanity and civilization. These two strand struggle for clarity and dominance in modern history. In a sense, Hebrew Scriptures are one of the earliest texts to deal with the concentric circles of loyalty to a specific group extending outward to caring about each individual, irregardless of group membership, simply because s/he is a human being. This essay will argue that the Holocaust may be seen as the culmination of a Western cultural history and philosophy which persisted in viewing Jews as alien and foreign, outside of nature and European nationalism. It will indicate several biblical and rabbinic sources regarding the centrality of the land

---

Great Powers, the Jews, and International Minority Protection, 1878-1938 (Cambridge: Cambridge University Press, 2006).

[5] The early name for the land, Judea/Israel, was changed to Palestine when conquered by the Romans in 63 BCE. Israel became the official name of the land with the establishment of the modern state in 1948 (following the partition of Palestine on Nov. 29, 1947, by the United Nations).

[6] See Joseph B. Soloveitchik, "Kol Dodi Dofek" (Hebrew). He calls the covenant of fate, *brit goral* and the covenant of destiny, *brit yeud.* This essay is translated by Lawrence Kaplan as "The Voice of my Beloved Knocketh," in Bernard H. Rosenberg and Fred Neuman, eds., *Theological and Halakhic Relflections on the Holocaust* (Jersey City, NJ: Ktav, 1992) and was first published in 1956. For an analysis, see Rochelle L. Millen, "Like Pebbles on the Seashore: J.B. Soloveitchik on Suffering" in David Patterson and John K. Roth, eds., *Fire in the Ashes: God, Evil and the Holocaust* (Seattle, WA: University of Washington Press, 2005).

[7] Hans Kohn, "Introduction," *Prophets and Peoples: Studies in Nineteenth Century Nationalism* (New York, NY: Macmillan, 1952).

of Israel and analyze the comments of the medieval thinker Moses Nahmanides on some of the cited passages, also touching upon Christian views of the relation of Jews to the land of Israel. Then it will discuss the nineteenth century Zionist writers, Moses Hess and Leon Pinsker, as well as the twentieth century thinker Ahad Ha'am. Each must be viewed within the context of Kantian and Hegelian thought about Jews and Judaism, which had a profound influence on the rise of National Socialism. The emancipation of Jews in European countries in the nineteenth century, stemming from the Enlightenment, camouflaged an animus rooted in anti-Judaism and strengthened by modern nationalism. Only when historical circumstances led to the establishment of Israel in 1948, when land and nature once more became part of the context of Jewish living, did the notion of the Jew, cursed by Christianity to wander eternally, began to diminish. The post-Holocaust founding of the modern state of Israel is closely tied to the modern recognition of Jewish peoplehood as essential to Judaism.

BIBLICAL AND RABBINIC SOURCES

The land of Israel is integrally related to the covenant[8] between God and the Jewish people. One way of understanding Torah, or the Hebrew Scriptures, is that it is not only a narrative, law, poetry, and history, but also a political constitution: how is this people to live as a community of individuals guided by a morality and ethos that protect them while reaching out to the stranger? The underlying assumption is that the obligation "to keep the way of the Lord" (Genesis 18:19) is bound up with creating a political entity that manifests justice and righteousness internally, within one's group, as well as externally, with those outside that group. The land is a focus for ethical statutes, an arena for ethical behavior. This is exemplified in Leviticus 19:9-10:

> And when you reap the harvest of your land, you shall not wholly reap the corner of your field, neither shall you gather the gleaning of your harvest. And you shall not glean your vineyard, neither shall you gather the fallen fruit of your vineyard; you shall leave them for the poor and the stranger; I am the Lord your God.

---

[8] On the concept of covenant, see Michael Walzer, Menachem Lorberbaum, Noam J. Zohar, eds., *The Jewish Political Tradition: Authority* (New Haven, CT: Yale University Press, 2000), Chapter I.

Both one's own community ('the poor') and those outside it ('the stranger') are addressed in these verses. So intertwined are the land and the ethical requirements that the text states the land itself *responds* (in whatever ways we understand providence) to the actions of the people, as found in Leviticus 20:22: "You shall therefore keep all my statutes (*hukotai*) and ordinances (*mishpatai*) and do them, that the land to which I bring you to dwell not vomit you out." Land (Israel), God (Torah), and people (Israel) are connected. The land is inimical to those who disregard the ethical statutes (Torah). Nahmanides's[9] commentary on Leviticus helps to explain this through delineating the concept of national differentiation. Each people has developed its own culture and history, he writes, influenced by geography, language, and other environmental factors. But among peoples, the national ideals are generally separate from the ethical and religious; the latter are, in a way, superimposed upon the national culture. There is no specific ethic, for example, in being an Italian or a Swede. But in the case of the people Israel, the religious and ethical converge with the national; each is integral to the other. This convergence is thus most fully realized in the land of Israel where ethical strictures and moral/religious ideals can be actualized under conditions of political autonomy. A sovereign political community must struggle with issues otherwise dormant in its culture: power, immigration, poverty, governance. Outside of the land of Israel, according to Nahmanides, too many foreign influences are active, liable to distract the people from their primary objective: the cultivation of holiness. Indeed, the comments of Nahmanides on the land are central to his commentary on Leviticus, chapters 18-20.[10] These chapters are suffused with the command to "be holy," to emulate God, as God is holy. Thus a profound characterization is made of the human being: s/he has the capacity to attain holiness–indeed, to be Godlike.

The holiness of which human beings are capable is related to the land; the land itself "will vomit out all those who defile it and will not tolerate worshippers of idols, nor those who practise [*sic*] immorality."[11] Why, asks Nahmanides, should

---

[9] Moses Nahmanides (1194-1270), also known by the acronym, Ramban (i.e. Rabbi Moses ben Nahman), was the Chief Rabbi of Catalonia and died in Israel after being exiled following the Barcelona Disputation of 1263. A noted physician and kabbalist, he authored a commentary on the Torah and on the Babylonian Talmud.

[10] See *Ramban (Nachmanides): Commentary on the Torah, Leviticus*, translated and annotated by Charles Chavel (New York, NJ: Shilo, 1974), pp. 268-285.

[11] *Ramban*, p. 269.

the Torah draw a connection between forbidden sexual relationships and the land? Indeed, "sexual relationships affect personal conduct and do not depend on the Land, [...] so why should the Land be affected by these personal immoral acts?" Nahmanides responds that the people Israel is different from other nations in that it has been given the land of Israel through which it will "be dedicated to His Name."[12] The land is a unique instance of – an opportunity for – ethical perfection. It represents what the nineteenth century would call nationalism, but aligned with the religious ideal of holiness. Nahmanides expounds similarly upon Deuteronomy 11:21, which contains the commandments of phylacteries and *mezuzah*.[13] While these commandments are personal obligations and therefore applicable everywhere, Deuteronomy links them to the land. "[...] that your days be multiplied, and the days of your children, upon the land," states verse 11:21 regarding these commandments. Even though they are not like agricultural laws which can only be followed when living in Israel, from this passage, which links these *mitzvoth* to the land, is deduced that *all* the commandments are meant primarily for those living in the land. Those practiced and kept in the Diaspora incline one toward holiness, accustom one to the discipline. But among them there is the ambiance of the practice session or rehearsal: the full concert can only be played on the land itself.[14] Nahmanides's view of the centrality of the land leads to a fascinating – and perhaps surprising – conclusion. The aim of Torah is to create a people immersed in and disciplined toward holiness, and the land of Israel is a *sine qua non* for the realization of this goal. Therefore, he surmises, commandments, or *mitzvoth*, followed outside the land are in fact only provisional. While they have merit – both intrinsically and behaviourally – their primary value lies in maintaining a vitally alive consciousness of the task of the Jewish people. *Mitzvoth* performed in the Diaspora are preparatory for total Jewish life in its homeland. Thus are the land and soil, the raw materials of nature, incorporated into the religious and national aspects of Jewish peoplehood.

THE LAND OF ISRAEL IN CHRISTIAN THEOLOGY: A BRIEF OVERVIEW

Traditional Christian theology as formulated in the Gospels, the Letters of Paul, and the early Church fathers understood Judaism to have come to an end with the

[12] *Loc. Cit.*

[13] A *mezuzah* is a small case containing specific biblical verses affixed to the doors of one's home.

[14] See *Ramban* on Deuteronomy 11:18, Genesis 26:5, Leviticus 18:25.

destruction of the Second Temple by the Romans in 70 CE. The termination of Judaism – both the religion and its national aspect – was understood as punishment not only for failing to accept Jesus as messiah, but also for facilitating his death by crucifying him. According to these views, Jews should have disappeared from history. But the early Church needed to come to grips with the continued existence of Jews and Judaism, and eventually – much later – with the existence of an ongoing and vital rabbinic tradition. Alterations were made in the theology to accommodate the failure of Jews and Judaism to fade into oblivion. Accordingly, Jews were seen as eternally condemned – indeed, cursed – as was Cain, to wander the earth.[15] Their persistent existence was understood as necessary to fulfill Christian theological expectations. In the same way as Christian thinking had to make sense of Jewish presence, it also had to find meaning in the non-reappearance of Jesus as originally understood by the Gospelwriters. Thus developed the doctrine of the Second Coming: Jesus was to appear again and with his second coming the messianic age, as described by the Hebrew prophets and adapted into Christian theology, would be ushered in. The continued existence of the Jews was postulated as a necessary condition for the Second Coming. This ontological status ostensibly protected Jews from being murdered, but given their alleged evils, it was acceptable to humiliate Jews in the ways promulgated as permissible by Church law. In 1205, therefore, Pope Innocent III described in a letter to the King of France the doctrine that Jews must survive in European (read 'Christian') history, but only in perpetual servitude. Christians cannot murder Jews, but may, with impunity – indeed with the encouragement of Church law – humiliate them. Innocent writes:

> The Lord made Cain a wanderer and a fugitive over the earth, but set a mark upon him [...] lest anyone finding him should slay him. Thus the Jews, against whom the blood of Jesus Christ calls out, although they ought not to be killed, lest the Christian people forget the Divine Law, yet as wanderers ought they to remain upon the earth, until their countenance be filled with shame and they seek the name of Jesus Christ, the Lord [...] [they] ought [...] be forced into the

---

[15] On the wandering Jew legend, see Alan Dundes and Galit Hasan-Roken, eds. *The Wandering Jew: Essays in the Interpretation of a Christian Legend* (Bloomington, IN: Indiana University Press, 1986) and the modern German novel by Stefan Heym, *The Wandering Jew,* (Evanston, IL: Northwestern University Press, 1999).

> servitude of which they made themselves deserving when they raised sacrilegious hands against Him.[16]

Innocent's idea was not a new one. Seven hundred years earlier, Augustine had written: "[...] the continued preservation of the Jews will be a proof to believing Christians of the subjection merited by those who [...] put the Lord to death."[17] And Augustine's words have precedent in Paul, Romans, especially chapters 9-11.

The status of Jews as theological "reluctant witnesses"[18] to the culminating revelation of Christianity; of Judaism as having been superseded by Christianity and therefore relegated to an artifact and an anachronism; as the destruction of the Second Temple and expulsion from Palestine by the Romans: these notions remained powerful in the pre-Nazi era, both in the Catholic and Protestant churches. One example is the statement of a Protestant clergyman from Berlin, Erich Klapproth, who courageously protested the riots of *Kristallnacht* in 1938. His comments are of interest because while he bravely protests (signing his name and indicating that he is a pastor) the violence, calling it unchristian and a "blot upon the good name of the Germans," he does so in terms that maintain theological doctrines going back centuries. He writes: "Israel is cursed and on trial because they were the first who rejected Christ," He also concurs with the German government that he cannot "disregard the sins that many members of the Jewish people have committed against our Fatherland, especially during the last decades [...] against the Jewish race."[19] All Jews were Christ-killers; the 7.6% of Germans who were Jewish in 1933[20] were undermining German society; and Jews were a 'race'. In his calling for Christians not to "den [y] ies (*sic)* the name

---

[16] Letter of Innocent III to the King of France, 1205, quoted in Robert M. Seltzer, *Jewish People, Jewish Thought: The Jewish Experience in History* (New York, NJ: Macmillan, 1980), p. 359.

[17] Augustine, *Reply to Faustus,* in *Disputation and Dialogue,* Frank Talmadge (ed.) (New York, NJ: Ktav, 1975), pp.27-32.

[18] See Stephen R. Haynes, *Reluctant Witnesses: Jews and the Christian Imagination* (Louisville, KY: Westminster John Knox, 1995).

[19] Quoted in Otto Dov Kulka, "Popular Christian Attitudes in the Third Reich" in Otto Dov Kulka and Paul Mendes-Flohr, eds., *Judaism and Christianity under the Impact of National Socialism: 1919-1945* (Jerusalem: Historical Society of Israel and Zalman Shazar Center for Jewish History, 1987), p. 257.

[20] The number and percentage of Jews in Germany from 1816-1939 is given in *Monumenta Judaica, Handbuch, Beitrage zu einer Geschichte der Juden in Deutschland.* My thanks to Timothy Bennett for locating this source.

of Christ"[21] by supporting the violence of November 1938, he uses the very allegations made by the Nazis and the churches.

One might lament the doctrine espoused by Innocent III, correctly pointing out that in another ten years he would convene the Fourth Lateran Council, and the promulgations against the Jews enacted then would be only a part of the Church's response to the various upheavals and challenges to church authority of the thirteenth century. But this would be to overlook – indeed, to ignore – the long-lasting effects of these doctrinal declarations. The proclamations regarding the status of Jews in Christian Europe hark back to the early Church Fathers and remain part of Christian historical consciousness, both positively and negatively, until today. Several excellent examples of the state of current Christian beliefs regarding Jews, despite the subtle, complex, and ongoing work of post-Holocaust Christian theologians, are given in the first chapter of Clark Williamson's study, *A Guest in the House of Israel*: *Post-Holocaust Church Theology.*[22] In this chapter, Williamson relates several incidents that occurred on Ash Wednesday of 1990, in which various pastors, in their sermons left intact or without explanation their language – or the language of biblical texts – which referred to Jews as 'hypocrites,' as those who 'hated Jesus' and who lost "all sense of proportion for justice and truth. They were so righteous in observing their religious law while they demonically pursued the death of Jesus." [23] That Jews were considered blinded by legalism and enthused to kill Jesus is viewed as doctrinally substantiated by the destruction of the Second Temple in 70 C.E. and the expulsion of Jews from Palestine, especially after the *Bar Kokhba* revolt of 132-135 C.E. Thus eviction from the land considered given to Jews in the biblical covenant, is seen as a proof of the demise of Judaism and a verification of replacement theology. Williamson's examples remind me of two experiences of my own.

The first occurred about four years ago when I was invited to be the Sunday morning speaker at a downtown Presbyterian church in Columbus, Ohio, where I live. I was one of several speakers in an educational series, programmed prior to morning religious services. The congregation was known to be well-educated and mostly upper middle class. I was therefore surprised to be asked to explain the 'legalism' of the Pharisees, a presumed parochial legalism contrasted with

---

[21] *Loc. Cit.*

[22] Clark M. Williamson, *A Guest in the House of Israel: Post-Holocaust Christian Theology* (Louisville, KY: Westminster/John Knox, 1993). Chapter I, pp. 1-26.

[23] Williamson, *A Guest in the House of Israel: Post-Holocaust Christian Theology*, pp. 2-3.

the alleged universal love of Christianity. Clearly, the attitudes remarked upon by Williamson were alive and well. The second episode was a year later, when I attended the chapel service at Wittenberg, because it was in memory of the late University Pastor, who had suddenly died in middle age. He and I had worked together and I was friendly with his widow. Although I rarely attend campus chapel worship except for *Yom Hashoah*, it seemed appropriate to pay tribute to him by being present at the service. But the lectionary reading included John 20:19-21. My face reddened and I thought: what am I doing in this place? What if I went to synagogue and a text claiming to be divine proclaimed that Christians are to be feared, since, as stated in John 8:44, they are children of the devil? Only one other person there, a colleague who teaches German, understood. Without my saying a word, he later said to me: "I would have said 'the authorities' and not 'the Jews.'"[24]
These events confirm what Williamson writes, that Christians

> project onto their Jewish neighbors attitudes and beliefs that they are taught concerning Jews in the biblical story. [...]As long as we uncritically refer to 'the Jews' hatred of Jesus' or to Jews as 'hypocrites,' churchgoers will hear these remarks not only about Jews of long ago but about their neighbors across the street.[25]

Deeply ingrained negative attitudes about Jews and Judaism are at the core of much of Christian theology. The self-understanding of Catholic and later Protestant theology, despite the differences between them, hinges on the narratives

---

[24] When I wrote the University Pastor to check on the passage read that day, this is what she wrote back. Her e-mail (April 29, 2007) confirms the increasing sensitivity of clergy regarding the possible anti-Judaism of various passages: "I m not sure whether you're addressing the problem with John's Gospel and its reading in public worship or not (regarding the Gospel's use of the phrase "the Jews"), but if so, it's even a more significant critique/concern because this Gospel is actually read every single year of the 3 year cycle on the 2nd Sunday of Easter. In addition, other passages from John are also read using similar language (as today's Gospel). I remember you saying once that you wondered why a church that felt free to translate the language into gender inclusive language couldn't also feel free to improve on John's Gospel's use of the phrase "the Jews," rather than translate it literally. I think that's absolutely on target, and what I've done now is just gotten into the habit of correcting the reading either while I read it or soon after in the sermon, but I think it's something that the National Council of Churches definitely needs to address."
[25] *Loc. Cit.*

described above: Jews bear responsibility for killing Jesus; were blinded to the truth he preached; were sunk in a decrepit system of spiritually empty legalism. Their Temple was therefore destroyed and they were evicted from the land upon which the covenant is based. This destruction and subsequent expulsion justify the contempt held toward Jews and Judaism in Christian history. Thus the land and what happened to it play a pivotal role in constructing the supersessionist theology, which even today lingers – and often remains powerful – among various Christian denominations.

It should be noted that the notion of catastrophe as punishment is evident in many of the Hebrew prophetic writings, as well as the liturgy and the corpus of the Talmud.[26] Christianity adapted that notion of explaining evil, one prevalent in the ancient world and in our days as well. But Christian doctrine added to the traditional prophetic versions of theodicy. It averred that the temple was destroyed and the Jews expelled because the Jewish people, in their blindness and stubbornness, refused to accept Jesus as messiah. Thus the Christian interpretation of the Jewish Diaspora after 70 C.E. becomes the foundation upon which replacement theology is based. Jews forced to leave the covenantal land of Israel is a bold and direct religious statement: Jews and Judaism have been superseded.[27]

THE EMANCIPATION OF EUROPEAN JEWS: LAND AND IDENTITY

The gradual emancipation of Jews from medieval European ghettos began in 1791 with the decree of citizenship of the French National Assembly.[28] Napoleon, carrying the ideals of the French Revolution outside of France, broke down the ghetto walls in Rome and elsewhere.

---

[26] See Zachary Braiterman, *(God) After Auschwitz: Tradition and Change in Post-Holocaust Jewish Thought* (Princeton, NJ: Princeton University Press, 1998), esp. Part I; See also David G. Roskies, *Against the Apocalypse: Responses to Catastrophe in Modern Jewish Culture* (Cambridge, MA: Harvard University Press, 1984).

[27] See Jeremy Cohen, *Christ Killers: The Jews and the Passion: From the Bible to the Big Screen* (New York, NJ: Oxford University Press, 2007).

[28] The debate and decree may be found in Paul Mendes-Flohr and Jehuda Reinharz, *The Jew in the Modern World: A Documentary History,* Second Edition (New York, NJ: Oxford University Press, 1995), pp. 114-117.

Yet how could Christian Europe accept as equal citizens those who had been segregated for more than 300 years,[29] regarded as devils, sources of evil, money-mongers, and purveyors of black magic and the blood libel? Much as the Emancipation Proclamation of 1863 would, a century later, require the Civil Rights Act of 1964 – and we in 2007 are still working on actualizing its ideals – the struggles for political emancipation of European Jews that began in the late eighteenth century would be frequently debated, with legal rulings both pro and con varying into the twentieth century.

No matter how German Jews felt, or how devoted to France Dreyfus was, antisemitism remained a potent undercurrent in European society. A detested religious group had been catapulted – in some cases eased, reluctantly or otherwise – into mainstream culture. Yet could Jews really fit into a Europe seized by the nationalism of the nineteenth century? Jews were a culture, a religious group, an ethnicity, a people in whose traditions yearning for the land of Israel was a central, if distant, dream. Could such a group be–or truly become–French or German?[30]

Underlying the many debates about civil and political rights for Jews during the nineteenth century there is a paradox, a contradiction played out both on the abstract, theoretical level and in the practical realm. German Romanticism, Hegelian philosophy, and political/social theorists–Christian and Jewish–articulate various aspects of the lengthy, ongoing, strange series of discussions which culminates in the Holocaust. With destruction of the Jewish presence in Europe, debate is no longer needed. In the nineteenth century political climate in Western Europe, and even in Russia (although there the circumstances are otherwise quite different), Jews are alien not only because they are not Christian, but also due to their lack of a homeland. When religious intolerance becomes philosophically unacceptable after the French Revolution, nationalistic anomalies move foreground. As France, German, Italy, and even forlorn Poland, in 1918, unify language, geography, religion, and land, Jews come to be seen as inauthentic; they don't belong. As land, patriotism, and nationalism become more prominent, the concept of the wandering Jew, the unattached outsider, the spirit without a body, broadens.

---

[29] The Church enactment regarding ghettos was promulgated at the 1215 fourth Lateran Council. But it was only with the Council of Trent in 1545-1463, in response to the Reformation, that many places established segregated quarters for their Jews.

[30] I am using France and Germany as the primary examples of nineteenth century conflicts regarding political emancipation of Jews.

Karl Gutskow, for instance, the left wing Hegelian thinker, wrote in 1838 that Ahasverus[31] "[the eternal wandering Jew] is a living corpse, a dead man who has not yet died." Judaism, like Ahaseverus, has outlived itself and is obsolescent.[32] While some thinkers advocated civil emancipation of the Jews, the powerful image of Jewish separateness made palpable this contradiction: Jews had to be incorporated into the nation-state not as Jews, but as full-fledged Germans or Frenchmen. One would emancipate Jews from the shackles of medieval political and religious restrictions not so that they could be transformed into 'true' Germans or Frenchmen or even Russians. Nearly every exchange concerning what came to be known as 'the Jewish Question' revolves around a permutation of this issue: Yes, Jews **should** be assimilated into German society; but if they are not part of the *volk*, if they have no connection to German soil, is the 'should' a goal that can be realized? Thus the same forces that lead to a reconsideration of the place of Jews in European society point to the reasons why such a rethinking would not succeed.

The paradoxical pull of these forces is evident in the various debates about Jews throughout the nineteenth century. Napoleon's questions compelled Jews to deny the national aspect of Judaism as a way to confirm their loyalty to France, the first state in the world that had offered Jews citizenship.[33] The exchanges among Moses Mendelssohn, Christian Wilhelm Von Dohm, and Johann David Michaelis[34] engage the same issues. Dohm's carefully constructed and judicious advocacy of the amelioration of the civic status of Jews not only espouses Enlightenment principles (The Jew is even more man than Jew), but also implies integration will make Jews more similar to the Christian population. Interestingly, Dohm advises that "The Jews should not be excluded from agriculture."[35] He states this not only as a suggestion of diversity of occupation, but also as a concrete connecting to the soil of the land of which they will be citizens. Dohm's statement represents a kind of incipient nationalism: land and loyalty are intertwined. Dohm's document is replete with Enlightenment language; he writes of

---

[31] See n° 15 above. In the Hebrew Bible, Ahaseverus is the Persian King who chooses the Jewish woman, Esther, to be his queen in the Book of Esther.

[32] Quoted in Susan E. Shapiro, "The Uncanny Jew: A Brief History of an Image," http://www.umass.edu/judaic/anniversaryvolume/articles/11-B4-Shapiro.pdf, 1.

[33] See the documents in Mendes-Flohr and Reinharz, *The Jew in the Modern World*, Second Edition, pp. 123-140.

[34] Mendes-Flohr and Reinharz, *The Jew in the Modern World*, pp. 28-49.

[35] Mendes-Flohr and Reinharz, *The Jew in the Modern World*, p. 34.

'equal right,' 'the rights of man,' and 'enlightened policy.' He admonishes Christians "to get rid of their prejudices and uncharitable opinions" and recommends that teachers of Christian doctrine emphasize "any man of any nation who does right finds favor with God."[36] One might say, from our twenty-first century perspective, that Dohm is moving towards pluralism. But Dohm's 1781 advocacy of civic rights for Jews yields mixed results. Michaelis's immediate response, written in 1782, is to dredge up a variety of anti-Jewish stereotypes together with social arguments. Equal citizenship for Jews "would gravely weaken the state," and "the Jews will always see the state as a temporary home, which they will leave in the hour of their greatest happiness to return to Palestine."[37] Writing from within a religious and political tradition which defined Jews as eternal wanderers and outsiders in European Christendom, Michaelis critiques Judaism for holding fast to the then distant dream of redemption on one's own land. He fails to connect this Jewish vision with the Christian conception of the Second Coming: each is an imagined future state representing religious and social completion, in which all will know God and peace will predominate. Yet the particularity of the Jewish dream – distant though it is – is seen as a threat to the particularity of the German Christian vision. This raises the contemporary question: can a specific nation with a dominant religious/moral outlook make room for other nations with different perspectives?

Michaelis's negative comments pave the way, especially in Germany and France, for the proliferation of analyses in the nineteenth century regarding the Jewish presence in Christian Europe. From Jewish efforts to alter – in some ways 'Christianize' – Jewish traditions to be more like 'them,' to the controversies in the 1832 and 1842 revolutions, Jews remain part of European culture, but also stand outside it. How else to understand the fears aroused when Jewish identity was blurred into German character traits? Assimilated Jews thought they had to erase 'Jewish' characteristics, but their very integration was seen as a sort of crime. Jews were feared when separated from Christians by religious law, and they were still feared–only perhaps more so–when no longer so readily identifiable. Arthur Schopenhauer suggest the solution to the Jewish question is conversion to Christianity; Jews assert their nationality "with unprecedented stubbornness [...] [but] live[s] parasitically on other nations." Jews and gentiles must marry each other so that Jews *qua* Jews will disappear and "the ghost will be

[36] Mendes-Flohr and Reinharz, *The Jew in the Modern World*, pp. 28-36.
[37] Mendes-Flohr and Reinharz, *The Jew in the Modern World*, pp. 42-43.

exorcized; Ahaseverus will be buried..."[38] Schopenhauer sought the disappearance of Jews, using language later heard in Nazi propaganda: Jews as parasites. Gutzkow, somewhat differently, sees emancipation as a means to fulfill the Christian curse that Jews should cease to be a people. "Emancipation," he writes, "would for the first time directly split the Jews apart from one another, who until now have simply been scattered, and would fulfill the curse that was foreseen by Christ, namely that the Jews should cease for all eternity to be a *people* [*sic*]."[39] Christian thinkers, including Kant, Hegel, Schleiermacher, Fichte, and von Arnim argued that Judaism was dead, and that the continued existence of Judaism was an anachronism the Enlightenment must work to erase. Kant's work, coming as it did at the end of the eighteenth century and setting the framework for Enlightenment thinking, led to a weakening of the earlier attempts in Christian theology to justify Jewish existence.[40] In the Enlightenment emphasis on universalism, the particularity of Jewish traditions was a relic. Kant avers that the "euthanasia of Judaism is pure moral religion, freed from all the statutory teachings," and that Judaism "must disappear" so there will be "only one shepherd and one flock."[41] Hegel's extensive theories of religion deeply influenced the array of issues surrounding 'the Jewish Question.' Hegel reverses the traditional doctrine of Jews as guilty of deicide by proclaiming that the 'death of Jesus in fact represents God's destruction of Judaism.' Judaism was destroyed in the very moment of the crucifixion. Hegel's dialectical model asserts that God was compelled to become incarnate in Judaism, as the universal must become particular. God decided to put the Jews to death through God's death on the cross.[42] Judaism remained a fossil, a ghost. Only through the demise of Judaism and the continuing dialectic towards Christianity could religious universality again be present. Yet this is the very time when nationalism, the specific and particular loyalties of specific and particular groups, was developing its various

---

[38] Arthur Schopenhauer, *Parerga and Paralipomena: Short Philosophical Essays* 2 (Oxford, NJ: Oxford university Press: 1974) pp. 261-264. Quoted in Susan E. Shapiro, "The Uncanny Jew," p. 5.
[39] Quoted in Susan E. Shapiro, "The Uncanny Jew," p. 5.
[40] See Immanuel Kant, *Religion within the Limits if Reason Alone* (New York: Harper and Row, 1960), esp. Book Three; Amy Newman, "The Death of Judaism in German Protestant Thought from Luther to Hegel," *Journal of the American Academy of Religion*, 61:3 (Autumn, 1993) 455-485. Also Emil Fackenheim, "Kant and Radical Evil", *University of Toronto Quaterly* (1953-1954), 339-353.
[41] Quoted in Amy Newman, "The Death of Judaism...," p. 461.
[42] Amy Newman, "The Death of Judaism...," p. 475.

forms. Kant and Hegel move conceptually toward universal reason and universal religion, but politically Europe is embracing the particular. The trajectory of Protestant theology, from Luther onward, demonstrates the transformation of beliefs and values from dogmatic doctrine into the social and political realms. Thus, "the locale of the death of Judaism was progressively transferred from the metaphysical to the social-historical arena."[43] The death of Judaism comes to be seen as a historical necessity and civic and political emancipation focus on Jewish "character." Is moral regeneration of Jews possible and does it not require abandonment of Judaism's national aspects? It is no surprise that these overt and covert currents in Western thinking, when allied with the rise of racial theory, ignite the conflagration that was the Holocaust.

SOME JEWISH RESPONSES TO EMANCIPATION: ZIONIST THEORIES, NATURE, AND THE LAND OF ISRAEL

The contradiction inherent in the debates about emancipation and 'the Jewish Question' is not lost among the Jews of Europe, despite widespread assimilation and the push to 'fit in.' For some, the rise of Reform Judaism provides a resolution. For others, the persistence of anti-Judaism in both its subtle and powerful nineteenth century manifestations, especially the Dreyfus Affair in France, the rise of racial antisemitismin in France and Germany, and the overt physical violence of pogroms in Russia signal a different development: that the Jewish people, even a century after the beginnings of the Enlightenment, remain foreign and alien in European culture. The only remedy is to establish one's own homeland, where one's ethnicity and identity are supported and nurtured. Moses Hess (1812-1875), Leon Pinsker (1821-1891) and Ahad Ha'am, the pen name of Asher Ginzberg (1856-1927) articulate the dilemma and formulate theories and suggestions as to how it might be resolved. Each recognizes the significance of the land as essential to Jewish self-definition. The land is important not only as a concept deeply embedded in biblical tradition, as evidenced in the first part of this essay, but also in its practical sense: land represents the human tie to nature, to nourishing the soil, to creating beauty and cultivating one's own crops. For centuries church law forbade Jews to own land or even to work it. Jews were cursed to be wanderers; land settlement and ownership connote permanence. Establishment of Jewish homeland would create a safe haven, a center for cul-

[43] Amy Newman, "The Death of Judaism...," p. 479.

tural creativity, an expression of the nationalism integral to Judaism and widespread in both Eastern and Western Europe. Jews would no longer be wanderers, foreigners who could never belong. In their writings, Hess, Pinsker, and Ahad Ha'am struggle with defining the contradiction they feel and with relating it to the national aspect of Judaism. In different ways and coming from varying theoretical frameworks and geographical locations, each advocates what comes to be known as Zionism: the return of Jews to the ancient homeland of the land of Israel.

Moses Hess, a German Jew and theoretician of socialism, knew both Karl Marx and Friedrich Engels. Initially he argued for integration of Jews into the universalistic socialist movement. Having fled Germany during the 1848 Revolution, he returned in the early 1860s, immediately sensing the rising antisemitism that accompanied the nationalism expressed in the movements toward the unification of Germany. In 1862, he published *Rome and Jerusalem*, in which the rise of Italian and German nationalism – especially with the rising intolerance of the latter – became the sparks leading him to advocate a Jewish national revival. His volume urges that Jews establish a Jewish socialist state in Palestine. This would confirm Jewish identity in the modern world, be in concert with the emerging national movements in Europe, and would be the most effective way to respond to rising and persistent antisemitism. In *Rome and Jerusalem,* Hess puts forth a viewpoint similar to that of Nahmanides discussed earlier: even while in exile, Jews must preserve their national identity through Judaism. At the same time, Jews must work towards restoring a Jewish state in Palestine. Hess even advises the eventual formation of a new Sanhedrin, or supreme Jewish court, which would update and modify Jewish law in accordance with the new status of political autonomy. Hess's 1862 book, deriving from his experiences with socialist theory and German antisemitism, as well as his early immersion in Jewish tradition, helps set the stage for what would become, thirty-five years later, the *First Zionist Congress*.[44]

Leon Pinsker, born in Russia, studied traditional Jewish texts with his father and then attended Odessa University, becoming a physician. Initially favoring assimilation and equal rights, as did Hess, he was led by the pogrom of 1871 in Odessa and the state-wide outbreaks in 1881 to alter his views. He came to believe that enlightenment and humanism could not lead to equality for Jews. Just

---

[44] For further analysis of Hess's views, se Shlomo Avineri, *Moses Hess: Prophet of Communism and Zionism* (New York: New York University Press, 1985).

two decades after Hess published *Rome and Jerusalem,* Pinsker wrote *Auto-Emancipation*, in which he exhorted fellow Jews to strive for a strong national consciousness and eventually, independence. In attempting to define the various strands of antisemitism, Pinsker pointed to the irrational, persistent phobic elements within it. He succinctly states that the Jew is everywhere – and to everyone – the Other: "to the living the Jew is a corpse, to the native a foreigner, to the homesteader a vagrant, to the proprietary a beggar, to the poor an exploiter and a millionaire, to the patriot a man without a country, for all a hated rival."[45] Pinsker came to see antisemitism as rooted in the persistent foreignness of Jews. Europe had always been Christian; for centuries, to be a European was to be a Christian. Even with the advent of Enlightenment ideas of toleration and equality, however, Jews remained outsiders. Jews would only truly be insider, accepted for who they are, in their own land.

The deep, psychic feelings against Jews, dominant in European culture could only be counteracted, according to Pinsker, by 'auto-emancipation.' Humanism, enlightenment, and civil emancipation would not succeed; they were rationally designed solutions to the embedded currents of Jew-hatred cultivated over centuries in Europe. Logic and reason could not extirpate rampant cultural and religiously validated images. At first Pinsker sought any place of safe retreat for European Jews. He came to understand, however, that such a retreat already existed in the land of Palestine. Pinsker wishes to arouse national consciousness among Jews, seeing the woes of Jews embodied in the lack of nationhood. In his extraordinary eulogy for Pinsker, Ahad Ha'am meticulously analyzes the path Pinsker trod to eventually establish the *Hovevei Zion*, or *Lovers of Zion* movement.[46] Pinsker articulated the need for a national will, a spirit that would find its regenerative and creative impulses within a national center.

Asher Ginzberg, known by his pen name, Ahad Ha'am, or *one of the people*, further develops the ideas of Pinsker. Ahad Ha'am is the great proponent of cultural Zionism. He does not seek a safe retreat, as did Pinsker originally, and as was central in the thinking of Theodore Herzl. For him, political Zionism would be efficacious only after a period of further nurturing of nationalist understanding and feeling. While this had already occurred to a large extent within the Rus-

---

[45] Leo Pinkster, *Auto Emancipation* (Maccabaean: Harvard University, 1906).

[46] See Ahad Ha'am, "An Open Letter to my Brethren: Pinsker and his Pamphlet, *Auto-Emancipation*." The full text can be found at http://www.zionism-israel.com/hdoc/achad_haam_letter_pinsker.htm.

sian Pale of Settlement,[47] it was less powerful a trend in Western Europe, where assimilation and emancipation were deemed reachable goals. Ahad Ha'am's cultural Zionism is complex, separating Jews and Judaism in an ideational split. Realizing that he could not yet proclaim the synthesis of both in Zionist theory, Ahad Ha'am proclaimed that "The land of Israel cannot be a safe retreat for Jews, but it can be, and it should be made, a safe retreat for Judaism."[48] What does this imply? For Ahad Ha'am, it means that in the return to Zion, Judaism will have "returned to its source." The land on which Judaism sprouted and came to maturity will once again nurtureand nourish the community, providing a physical framework for an otherwise lost spirit. In Palestine, there would exist a "'national, spiritual center' for Judaism, a center beloved of all the people, serving to unify the nation and fuse it into one *body* [*emphasis mine*]; a center for law and science, for language and literature, for physical labor and spiritual elevation; a miniature representation of what the Jewish people ought to be."[49] And Ahad Ha'am realizes his vision is a long-term, distant vision. "But," he writes, "no road, however long, may seem too long to the wanderer of thousands of years."[50]

CONCLUSION

That Jews share an ethnicity and sense of peoplehood is central to the Hebrew Scriptures and to the rabbinic traditions and liturgies which derive from it. Yet it is this shared national destiny which has contributed to the persistence of anti – Judaism and the rise and flowering of racial antisemitism, which culminated in the Shoah. Hindsight teaches us that the emancipation of Jews in Europe failed to erase –perhaps even to diminish – the profoundly ingrained notion of Jews as foreign, alien, and in some ways, despite their tiny number, responsible for vast societal evils.[51] Nineteenth century nationalism, constructed upon the foundation

---

[47] See Howard M. Sachar, *The Course of Modern Jewish History*, (Vintage: Revised Edition 1990), p. 323.

[48] Ahad "Ha'am, "An Open Letter....," p.7.

[49] *Loc. Cit.*

[50] Ahad Ha'am, "An Open Letter...," p. 8.

[51] See for instance, Steven E. Aschheim, *Culture and Catastrophe: German and Jewish Confrontations with National Socialism and Other Crises* (New York, NJ: New York University Press, 1996), especially Chapter 3, "'The Jew Within': The Myth of 'Judaization' in Germany," pp. 45-69.

of Enlightenment principles of toleration, equality, and the "rights of man," could not embrace – or even accept – the Jewish presence in its midst. Indeed, much of the nineteenth century may be understood as a history of the wavering of European Christians to realize liberal Enlightenment political principles and to decide how its Jews were to be treated. Would Jews be citizens, as in France, which erupted one hundred years after the granting of citizenship during the Dreyfus Affair? Would Jews be integral to Germany, even though over eight printings of Houston Chamberlain's 1899 vitriolic racist volume, *The Foundations of the Nineteenth Century*, were circulating in Germany by the first decade of the twentieth century?[52] Russia, where Jews were already pauperized by deliberate government policy, revived the Blood Libel in 1911 to detract attention from its crumbling empire. How could a Jew survive as a whole person, as free and equal, without unease and fear? Christian anti-Judaism allied with nationalism and racial antisemitism is a potent mix, and by 1945 it succeeded in making Europe nearly completely *Judenrein*. The return to nature and the land of Israel, made official in 1947 by the United Nations, affirms – in Buber's words – "[...] the Jew as the bearer and the beginning of nationhood."[53]

[52] William Shirer, *The Rise and Fall of the Third Reich* (New York, NJ: Simon and Schuster, 1985), p. 107.
[53] See n° 1.

In Response to Rochelle L. Millen

Peter Haas

Rochelle L. Millen illustrates how people have related to *nature* by looking specifically at the Jewish attachment to the Land of Israel. In her essay, she traces this attachment through roughly two millennia of history, from biblical times to the contemporary State of Israel. My response looks at the various examples that Millen provides, and argues that while they are all Jewish and all related to the land, they reveal not one, but many types of relationships. In other words, there is not a single or monolithic Jewish view regarding the Land of Israel. As I shall argue below, there are different types of attachment to be found in the literature and each reflects a very different Jewish notion of nature and the role of nature in history.

For the writers of the Bible, of course, the land was of very specific importance. It represented the place that the divine had chosen for his Name to dwell, to use the Deuteronomic phrase. This decision on the part of the Divine meant, according to biblical thinking, that the land itself had to undergo some transformation in order to be able to tolerate the divine Presence. Part of this transformation involved clearing the land of its 'pagan' inhabitants, along with their various altars and high places. In their stead came the covenanted People of Israel, who were to inhabit the land in accordance to the divine will. This re-settlement was to provide the substructure for the dwelling of the Divine name. The altars and high places, in turn, were replaced by the appointment of Jerusalem as a holy city, and the construction of a single holy temple within that city.

There is a clear sense articulated in many places in the Biblical books, especially Ezekiel, that the Israelite settlement was constructed so as to create a series of nested areas. In the center, with the highest degree of sacrality, was the altar of the Temple, that precise location from which the offerings went from earth to heaven. Beyond the altar were areas of ever decreasing holiness: the priestly courtyard, then the area that could be entered by all ritually clean male Israelites, then the 'women's courtyard' into which any ritually clean Israelite could enter, then the general city, then the boundaries of the Land of Israel, and finally all the rest of 'the nations'. Each of these areas was governed by its own rules so that its particular level of sanctity could be maintained. One such set of rules governed who could enter at all, as we have just seen. Another example is provided by the disposition of dead animals. Thus, for example, only certain animals could be killed in the altar area, and then only in certain tightly prescribed ritual ways. Meat from the sacrifices was occasionally given to the donor, but could only be eaten within the walls of Jerusalem. Outside the city, the killing of animals was much less controlled, although the blood had to be buried in all cases. Beyond the Land, no such rules applied at all. In each case, the status of the land both determined, and was affected by, human activity.

This notion of the potential sacrality of nature took on an entirely different meaning in the diaspora Judaism of the Greco-Roman period. For rabbinic Judaism, the exile from the land and the ultimate promise of the return to it, took on an eschatological meaning. The Jewish exile from the Land was taken to reflect a sort of divine alienation and the promised return would be a sign of redemption. In this view, the overrunning of the Land by strangers was not only a national misfortune, but had cosmic implications as well. It meant that the dwelling of the divine in the Holy Land was rendered impossible. This removal of the divine from the land also represented, of course, a larger threat to the very stability of Creation.

On this view, the return to the land was of utmost importance, not just for the People of Israel, as in the biblical view, but for the world altogether. It meant that the divine Presence would once again return to reside fully in Creation, and by that very fact induce (or constitute?) the final end of history. Nahmanides, and others, expressed this view by saying that of the various commandments given to the People of Israel to shape a holy life, some could only be done in the Land. The larger implication, it seems to me, is that it was impossible for humankind altogether to enjoy redemption while Jews were in a diaspora situation. Thus the return to the Land would be a completion, as it were, of the divine intention to effect the sacralization of all Creation. To be sure there were plenty of questions in the tradition as to how and when such a return might occur. But there was almost total unanimity that such a return would occur and that it would have a cosmic impact. The redemption of nature, or maybe more accurately, the release of the holy potential in the Land, was central to the whole drama of the universe.

This view received maybe one of its clearest overt expressions in the mythology of the sixteenth century *Lurianic Kabbalah*. In this literature, the original act of creation resulted in a divine 'accident' in which the divine was 'shattered' and sparks or shards of the divine became embedded in the natural world. The divine could be restored only if the sparks could be freed from their entrapment in nature, and this could happen, in turn, only through the purposeful fulfillment of divine commandments of how to conduct oneself in the natural world. It was from then on, in and through nature, specifically the Land of Israel, that the *tikkun*, that is the repair of the world and its final completion and redemption, could occur.

The modern world saw the enchantment of nature in two different ways. One was the notion of the Land of Israel as sacred because it meant the physical survival of the Jewish nation. According to this particular view of Zionism, the Land of Israel (*Zion*) represented the roots of the Jewish people, and so the only soil out of which their (secular) national reawakening could take place. To be sure, not all Zionists saw the nature of the Land of Israel in this way. Theodore Herzl, the founder of modern Zionism, famously suggested in all seriousness that the Zionist Congress considers establishing a Jewish homeland in central Africa. But for the vast bulk of Jews, if there was going to be a national revival, then it could only be conceivable on the one piece of nature to which the Jewish people had a 'natural' attachment, namely the Biblical homeland. On this view, the land was sacred, and working it had a holy quality, because it promised communal survival. In contrast to the medieval view, now working the land, planting trees, draining swamps, became 'sacred' not because doing such deeds were fulfilling divine commands. Rather, such activities were sacred, in a Durkheimian sense, precisely because they were fulfilling the needs of a community to survive and reconstruct itself. There were of course many 'religious' Zionists who came to the Land of Israel for purely religious reasons, as noted above, so as to be able to fulfill even those commandments which related only to the Land of Israel. But for many of the Zionist movements, the sacredness of the Land was something else entirely. It was only in this place that the secular nation of Israel could legitimately and naturally be replanted.

The second modern Jewish relation to the land is that of the average, secular Israeli Jew. For secular Israelis the natural landscape is sacred not because it represents the divine Presence, or because it allows full observance of Jewish law, or because it allows for a return. The land is 'sacred' because it is literally, home. As such, however, it is both a place of comfort and a domain of peril. Israelis relate to their natural landscape both as a place of work, rest and recreation on the one hand, and as an arena of warfare and terrorist attacks, on the other. For contemporary Israelis this natural world is both friendly and alien. The pile of stone ruins can bespeak both the ancient roots of the Jewish people in the Land, but also the recent history of prior inhabitants who turned into refugees. Hills can represent points of military power and hence safety and security, or points of menace from which missiles can be launched at the valleys below. The very lack of natural boundaries suggests a certain lack of closure, and an open-endedness to the political future of the area. So Israelis today, dealing with their natural surroundings, see reflected their own existential situation in the landscape.

My point is that these four senses of the Land – biblical, rabbinic, Zionist and Israeli – represent four distinct orientations to the natural world of the Land of Israel. While these four are clearly linked historically and in people's minds, they do represent, I think, different relationships that Jews and Judaism have established with the Land and with nature.

In Response to Rochelle L. Millen

John K. Roth

> You shall therefore keep all my statutes and ordinances and do them, that the land to which I bring you to dwell not vomit you out.
>
> Leviticus 20:22

Rochelle Millen's post-Holocaust reflections on "Land, Nature, and Judaism" raise four main questions that I hope she will address. Stating them effectively requires establishing a context, for the issues arise not only from her explicit statements and governing perspectives, but also from her essay's silence, which is particularly notable in a book that, among other things, explores how Holocaust-related studies have implications for ecological ethics and theology. A verse she quotes from, Leviticus 20, which also serves as the epigraph for my remarks, links these aspects.

Leviticus 20:22 epitomizes an outlook that decisively informs Judaism and fundamentally influences Jewish life. Key dimensions of that worldview appear in other traditions, too; they are often linked to nationalism. Whether the worldview is Jewish or non-Jewish, the crucial idea is that a people and a part of the earth belong together and that this relationship is divinely ordained. Judaism stresses a covenantal understanding of the people-land-divinity bond. That bond requires the people's devoted observance of divine commandments. Failure to honor these statutes and ordinances, as Leviticus calls them, does not take place with impunity. Instead, the divinely given land in which the people dwell will vomit that people out.

Vomit – the word refers to nothing pleasant. As a verb, it denotes stomach-based retching, spewing, and disgorging that are signs of disorienting and uncontrolled sickness. As a noun, vomit identifies a stinking, nauseating mess. In any of its forms, vomit does not appear out of the blue. Something causes it. Leviticus 20 indicates that the behavior of a people who have received divinely given land could make that land retch and spew. Such vomit would disgorge the very people who cause it. Leviticus 20:22 is harsh, even if accompanied by the recognition that vomit may signify purging and be a step toward restored health.

Millen invokes Leviticus 20:22 to underscore two themes. Primarily, she stresses that the text shows how "the land is a focus for ethical statutes, an arena for ethical behavior." Underlying that emphasis is an even more fundamental point: Leviticus 20:22 reflects the view, as Millen puts it, that "the land of Israel is integrally related to the covenant between God and the Jewish people." Or, as she also says, "Judaism differs from other religious traditions in that its full realization, according to Hebrew Scripture and the rabbinic texts which interpret Scripture, occurs in a specific geographic location: the land of Israel."

In traditional Judaism and in Millen's explicit stance as well, the warning in Leviticus 20:22 is the counterpoint to ethical and national idealism. "In the case of the people Israel," she writes, "the religious and ethical converge with the national; each is integral to the other. This convergence is thus most fully realized in the land of Israel where ethical strictures and moral/religious ideals can be actualized under conditions of political autonomy." Some version of that viewpoint finds expression in most, if not all, Jewish hopes for a homeland. "The land is important," Millen indicates, "not only as a concept deeply embedded in biblical tradition [...], but also in its practical sense: land represents the human tie to nature, to nourishing the soil, to creating beauty and cultivating one's own crops." She adds one more crucial element to the motif of ethical and national idealism: According to modern Zionist convictions – pre- and post-Holocaust – the "establishment of Jewish homeland would create a safe haven, a center for cultural creativity, an expression of the nationalism integral to Judaism and widespread in both Eastern and Western Europe."

Millen quotes Martin Buber's 1916 description of "the Jew as the bearer and beginning of nationhood." That idea was strongly related to Buber's beliefs, in her words, about "the profound connection between the land of Israel and the religion of Judaism at a time when issues of civic emancipation of Jews in Germany seemed smoother and less problematic." Those issues did not stay smoother and less problematic for long. Nor did a host of others, including the hope that the State of Israel would be a safe haven, and thus the issues that I hope Millen will address come to the fore.

My first question, which is rooted in my impression that Millen's reflections have a predominantly pre-twentyfirst century focus, can be stated as follows: What might she say to remove her essay's disturbing silence about a critical reality that could render her concerns insignificant? I apologize for the fact that this query has some harshness akin to that of Leviticus 20:22, but my intention is to cut to the chase as clearly and provocatively as possible. The silence I have in mind is that, with only minor and mostly implicit exceptions, Millen's essay proceeds as though the threats of global warming and climate change are not upon us with a vengeance. Especially in a post-Holocaust world, Leviticus 20:22 has its problems, as I will point out briefly below, but on a basic point that ancient text may get things right, and I am surprised that Millen does make not much of it. The basic point is that humankind's stewardship of the earth has long been so beset by ignorance, greed, shortsightedness, and weakness of will that the land, figuratively, if not in some actual way, may vomit us out. It is no exaggeration to say that human civilization, at least as we know it, is endangered because we abuse nature to such an extent that it cannot sustain the culture we have created, including its religious, philosophical, political, economic, scientific, and national traditions. True, emphasizing specifically Jewish ones, Millen speaks of human ties to nature that are nourishing, creative, and beautiful. Nevertheless, as I read her chapter in times of reckless consumption of natural resources and careless devastation of nature itself, which is actually self-destruction because human life is part of the natural world, I yearn for her to relate her post-Holocaust reflections on land, nature, and Judaism to humanity's ecological calamities.

If Millen goes in those directions, two further questions loom large. First, how should nationalism be evaluated? Much of Millen's essay is a critique of nationalisms – religious and secular – that marginalized, excluded, and ultimately annihilated Jews. At the same time, modern Zionist responses advanced the aim of a Jewish state that would occupy at least some of the biblical land of Israel and be positively different from other national entities. What is missing from Millen's account, however, is an exploration of the relationship between nationalism and the ecological dilemmas that ensnare us. Arguably, the threats of the latter are a consequence, however unintended, of nationalism – particularly the modern nation-state – and its ways of wedding peoples to particular parts of the earth. Nationalism, of course, is by no means the only cause of these problems, but it deserves consideration as one of the key contributors to them.

Particularly since the eighteenth century, the identities of people, of nations, are linked to nation-states. In one way or another, moreover, national interests are tied to the occupation, control, and use of nature, not only but particularly land. Those ambitions and initiatives typically entail violence, at least the threat of it, and often the despoiling of the natural world in the process. 'Country first' is not just a political campaign slogan; it is a life-and-death commitment for many people all around the world. Unfortunately, human beings may be pursuing national interests in ways that will make the land, as Leviticus suggests, vomit them out – persons, nations, and interests alike. Buber may have correctly identified the Jew as 'the bearer and beginning of nationhood,' but nationhood has been a mixed blessing at best. The Holocaust, as well as environmental stress, bears witness to that, for absent nationalism Nazi genocide would have scarcely been conceivable. I believe that Millen agrees, but can she relate that agreement more explicitly to the ecological struggle that humanity faces?

She might do so by showing how nationalism could be part of the solution for our ecological dilemmas. So, especially given her favorable view of the Jewish land-people-divinity bond, my third question is what would happen if she turned her attention in that direction? Millen could turn, for example, to the high level of ethical and national idealism that has infused Judaism and hopes for a Jewish state in particular. That idealism might help to show how we could mend our ways ecologically. She could also turn to the fact that national life is increasingly globalized and interdependent. The first turn could be the basis for an argument that nationalism need not be gripped by narrow visions of national interest; nationalism could embrace a sound ecological consciousness and conscience. The second turn could be the basis for an argument that national interest can be rational only if it is interest rightly understood, which is to say that national interests must recognize that the earth is our mutual home and that the fouling of one's own nest will not be beneficial for oneself, let alone for future generations. Directed in those ways, at least some of the negative features of nationalism could be restrained, and a more rational pursuit of national interest might have salutary effects for our natural environment.

If Millen takes those turns, however, there are problems with both. Some nations do better than others, but a fair empirical judgment is that every nation falls far short of its highest and best ideals. The shortfall, morever, is presently enough to keep ecological dilemmas brewing. Far from successfully warning us about that destiny, globalization seems to be doing rather little, at least for now, to reduce the wasting of the natural world. Globalization may hold crucial keys for reversing that situation, but the value of that promissory note remains to be seen. We need Millen's insights about the realistic turns that can be taken as we face a slow but seemingly sure deterioration of humanity's natural environment, a process whose destructiveness in the long run could even eclipse the waste that the Holocaust and other genocides have produced.

As Millen takes stock of the realistic turns that can be taken, a fourth question arises: What can religion do best to help us cope with the environmental threats that we are inflicting upon ourselves? In the context of Millen's essay, that question pertains especially to Judaism and Christianity, the two traditions that primarily occupy her attention. With regard to those considerations, what role, if any, can be played by the theology contained in Leviticus 20:22? That text is steeped in the view that the world is God's creation and that God is essentially and providentially involved in human history. One may be able to identify with the Leviticus theme that human greed and folly can result in the earth's spewing us out, but in a post-Holocaust world, is it credible to think that Leviticus got it right about God and the relationship that text affirms about history-nature-divinity, a relationship in which God is profoundly and ethically engaged with the historical process and its interaction with nature? Can the harshness of Leviticus be a credible, if not entirely welcome, alternative to the despair that may accompany a creation devoid of divinity and the combination of judgment and hope that a bond among nature, history, and divinity might involve?

In Reply to My Respondents

Rochelle L. Millen

I much appreciate Peter Haas' further amplification of what I have written regarding land, nature, and Judaism. His claim that the relationship described is in actuality not a single one, but takes at least four different forms, is indeed the case. The overarching connections among land, nature, and Judaism has yielded nuanced distinctions in the history of the tradition. That the holiness of the land becomes a venue for redemption of the world, "to the very stability of the Creation," in Haas' words, is corroborated by several biblical passages, among them Isaiah 56:6-7 and I Kings, 8:41-43. As these verses make clear, the "foreigner" is welcomed in God's Temple, which is "a house of prayer for all peoples" (Isaiah, 56:7). The more people who acknowledge God as Creator, the more the world moves closer to redemption.

The biblical, rabbinical, mystical, and modern meanings ascribed to the intertwining of land, nature, and Jewish theology indicate the syncretism intrinsic to Jewish thought. In the four senses explicated by Haas, the holiness of the land to the Jewish people remains pivotal, whether in eschatological, Durkheimian or political terms; his additional explanations further elucidate the fundamental concepts established in my overview.

John Roth poses four fascinating questions as commentary on Leviticus 20:22. My interpretation of this verse points to its ethical and national ramifications, both pre-Holocaust – i.e. nourishing the soil, creating beauty, and sustaining one's community – and additionally, post-Holocaust, creating a safe haven and center for cultural creativity. My response to the first question, "the disturbing silence" regarding "the threats of global warming," is that this significant issue is indeed addressed, although by implication and not explicitly. For the notion of wise, careful stewardship of the land is latent in many commandments of the Bible, regarding not only the land of Israel but also any land upon which one may dwell. Thus when Scripture links ethics and moral behavior to worthiness of inhabiting the land, part of that merit is care of the land itself. Even the very first earth creature is commanded not only to conquer and subdue the earth (Genesis 1:28), but also to till, cultivate, tend, and nurture it (Genesis 2:15). Indeed, the earth creature is commanded both to stride forward by creating and dominating, and to retreat as well, to allow the natural growth and regeneration embedded within the physical universe created by God. These dual aspects of human connection to the earth is made explicit for the Jewish people by the requirements of the sabbatical and the Jubilee years (as in Leviticus 25), periods of time when the land must lie fallow, nourished by its own intrinsic powers, and thus restored to its necessary richness. Human beings must relinquish their push toward domination and utilization of the land, acknowledging the land's autonomy as God's creation and its need for revitalization. The same theology which governs the Sabbath day for people, i.e. recognition of God as ultimate Creator, undergirds the sabbatical of the land.

John's second query asks how nationalism and ecological concerns can aid one another rather than lead to destruction. Certainly, German nationalism in the years of Nazism was characterized by a rapacious ethic; when nation becomes all and is subverted to any notion of universal ethics, abuse of natural (as well as cultural) resources becomes endemic. But such abuse is not a *sine qua non* of nationalism. It is perhaps idealistic to claim that nation states can embody agreement and cooperation, but such idealism is not without precedent.. There has been sharing of water resources, agreements about chemicals, mining and fishing rights. As in any human enterprise, selfishness and greed can impede the success of such treaties. But even as nationalism accords legitimacy to diverse cultural groups, it need not imply that self-interest trump ethics. As in individual, communal, and societal interactions, the vision of a greater good can inspire conservation and cultivation. Recycling of paper and plastic, energy efficient appliances, and water saving gadgets are all examples of our attempts to reverse the destructive effects of earlier policies. National interests that violate the right of the natural world to exist safely cannot be deemed acceptable in the court of world opinion. However, as John notes, the deterioration of humanity's natural environment continues, despite the best efforts of some. I wish I had more penetrating insights to offer about what "realistic turns" could be taken to increase the minimal and belated, if growing efforts to alter the moves toward contamination and diminishing of our formerly rich resources.

John's fourth question is especially incisive and poignant, as it challenges us to reconsider and reevaluate the history-nature-divinity nexus at the core of Leviticus 20:22. Is it credible, he inquires, to have faith in the judgment and promise implicit in the text? In the post-Holocaust world, what allows us not to despair? The God-nature-history bond is the very foundation of Judaism, as well as Christianity and Islam. While deeply cognizant of its paradoxical aspects, I can relinquish neither individual/societal responsibility nor hope, even post-Shoah. I can only affirm Mishnah Avot, 2:20: "It is not your duty to complete the task, but you are not free to desist from it." For me, the task the Mishnah refers to is continuing to work toward a world in which evil diminishes. It is the only path pointing in the direction of mending the brokenness the Shoah brought about.

The Ashen Earth:

Jewish Reflections on our Relation to Nature in the Post-Holocaust Era

David Patterson

"We recognize that separating humanity from nature [...] leads to humankind's own destruction," said German botanist Ernst Lehmann in 1934. "Only through a reintegration of humanity into the whole of nature can our people be made stronger. That is the fundamental point of the biological tasks of our age. Humankind alone is no longer the focus of thought, but rather life as a whole [...]. This striving toward connectedness with the totality of life, with nature itself, a nature into which we are born, is the deepest meaning and the true essence of National Socialist thought."[1] Thus the man who saw National Socialism as 'politically applied biology' stated the Nazi position on the physical and spiritual relation between the *Volk* and its natural environment. If 'life as a whole' is now the center of consciousness, then the Jews are not a mere ethnic, social, political, or economic threat – they are a *cosmic* threat to 'the totality of life,' and their eradication must be total.

A year after Lehmann made his statement, the *German physicians' association* asserted, "The comparison between Jews and the tuberculosis bacilli is a telling one. Almost all people harbor TB bacilli, almost all nations on earth harbor the Jews – a chronic infection, which it is difficult to cure. Just as the human body does not absorb the TB germs into its general organism, so a natural, homogeneous society cannot absorb the Jews into its organic association."[2] Both Lehmann's statement and the assertion from the German physicians' association were made in the contexts of a famous remark made in 1930 by Nazi race theorist Richard Walther Darré, who declared that the challenge facing the German people was the restoration of "the unity of blood and soil"[3] – hence the Nazi

---

[1] Ernst Lehmann, *Biologischer Wille: Wege und Ziele biologischer Arbeit im neuen Reich* (Munich: Zentralverlag der NSDAP, 1934), pp. 10-11.

[2] Quoted in Hayim. H. Ben-Sasson, ed., *A History of the Jewish People* (Cambridge, MA: Harvard University Press, 1976), p. 1019.

[3] Richard Walther Darré, *Um Blut und Boden: Reden und Aufsätze* (Munich: Zentralverlag der NSDAP, 1940), p. 28.

slogan *Blut und Boden*, "Blood and Soil," and the definitive connection between the purity of the race and the purity of the earth. What affects one affects the other, so that the Jewish pathogen endangers not only the body and soul of the German people but also the very earth beneath their feet. The implication? Like the tuberculosis bacilli, the Jews must be completely exterminated – for the sake of the *Reich*, for the sake of the *Volk*, for the sake of the earth itself.

From a Nazi standpoint, the ties that bind together blood, soil, and *Volk* are both physical and metaphysical. For the Nazi, race is first philosophy. To say that National Socialism is rooted in the German soil is to say that it is rooted in the German soul. Its philosophy is steeped in blood and soil, and it opposes above all the philosophy that lurks in the blood of the Jew. Just so, the infamous Nazi ideologist Alfred Rosenberg declared that the *Blut und Boden* of the Aryan race had been poisoned not merely by Jewish blood but by Judaism, for the -ism is in the blood.[4] All Jews are prone to think 'talmudically,' he insisted, "whether they are atheistic Bourse-speculators, religious fanatics, or Talmudic Jews of the cloth."[5] And so the Aryan murderers purged their blood of Jewish blood and covered the soil with Jewish ashes. In the process they issued prohibitions against Jewish prayers and ritual baths, desecrated Torah scrolls and cemeteries, and planned their actions according to the Hebrew holy calendar. Yes, the target was not only the Jewish people but also the teaching and testimony of Jewish tradition, that the Nazis believed had emerged from Jewish blood. The Nazis, therefore, were no mere racists: they embraced a worldview, a *Weltanschauung*, that was antithetical to the Jewish view and that could not inhabit the same earth that the Jews inhabited.

Because the Nazis set out to exterminate both the Jews and Judaism, in the post-Holocaust era we must turn to what was slated for destruction if we are to overcome the Nazi assault. Responding to the Shoah, we must think talmudically, if we are to overcome the pagan idolatry that characterizes Nazi thinking and that continues to characterize much of what passes for environmentalist thinking. To be a witness to those lives reduced to ashes is to be a witness to the teaching they represented by their very presence in the world. And part of that testimony lies in the connection between the absolute, Divine commandment not to murder (Exodus 20:13) and the absolute, Divine commandment not to despoil the earth and

---

[4] Alfred Rosenberg, *Race and Race History and Other Essays*, ed. Robert Pais (New York, NJ: Harper & Row, 1974), pp. 131-32.
[5] Ibid., p. 181.

its creatures; indeed, one of the Six Orders of the Talmud, *Zearim*, is devoted to our relation to the land that G-d has placed in our care.

The need to offer a Jewish response to Nazi environmentalism is greater than ever, as more and more of the world's environmentalist movements adopt a pagan, idolatrous discourse reminiscent of the Nazis.[6] Much of the modern environmentalist thinking devalues human life in the name of a nature that has become the ultimate value; once nature thus becomes an object of worship, anything and anyone can be sacrificed in the name of environmental purity. Such a view plays into the hands of the naturalistic idolatry that led the Nazis to prohibit the vivisection of animals in medical experiments, while permitting the same procedures to be performed on Jewish children. Of course, for one who sees nature or the environment as itself holy, there is no contradiction here. Where, then is nature situated in the Nazi scheme of things?

NATURE IN THE SCHEME OF NAZI IDOLATRY

Anyone who has seen Leni Riefenstahl's *Triumph of the Will*, the most infamous of the Nazi propaganda films, will recall the bucolic settings in which a vigorous and vibrant German *Volk* enjoys life in the midst of their natural German environment. Of course, the film's images of an Aryan humanity at home with nature are no accident. But what does this vision of the Nazi *Volk* living happily in the bosom of nature have to do with Nazi idolatry?

Much of the Nazi thinking about the relation between humanity and nature is rooted in the essay 'Man and Earth' (1913), published by the notorious anti-Semite Ludwig Klages in 1913.[7] Arguing for a life lived in harmony with the earth, Klages provided the Nazis with the view of a spiritual wholeness with nature that, in the words of Lucy Dawidowicz, "would restore them to primeval happiness, destroying the hostile milieu of urban industrial civilization that the Jewish conspiracy had foisted on them."[8] It is not that the Jews have no tie to the earth; Rosenberg, in fact, accused the Jews of being so earth-centered that they

---

[6] For a good discussion of this point, see Peter Staudenmaier and Janet Biehl, *Ecofascism: Lessons from the German Experience* (Oakland, CA: AK Press, 1995).

[7] See Ludwig Klages, *Sämtliche Werke*, Vol. 3 (Bonn: H. Bouvier, 1964), pp. 614-30.

[8] Lucy Dawidowicz, *The War against the Jews 1933-1945* (New York, NJ: Holt, Rinehart and Winston, 1975), pp. 61-62.

lacked a soul.[9] What characterizes the Nazis' pagan worship of nature is just the opposite of this earth-centeredness: it is a spiritualization of nature into the essence of the soul or of an invisible god, whose traces are manifest in flora and fauna. Indeed, more than a primeval happiness, the Aryan nation sought its primal essence in a pagan embrace of nature: to get back to nature is to get in touch with the soul. The soul does not emanate from G-d, filled with commandments, as the Jews maintain; no, it is in the Aryan *Blut und Boden*. Therefore the cosmopolitan, industrialist Jew, who contaminated the Aryan blood and poisoned the Aryan soil, imperiled the Aryan soul.

From a Nazi viewpoint, just as Judaism was in the blood of the Jew, so a certain conception of nature was in the blood of the Aryan. A 'feeling for nature' and a 'racially determined concept of nature,' wrote Bruno Thüring in 1936, characterize the "Nordic man, who strives to comprehend nature not only with his intellect, but also with his heart and soul and with his imagination."[10] Nature issues no commandment, but only a summon to majesty and imagination; indeed, defined solely by nature, the human being is limited only by the imagination. Thus what the Nazis ultimately did to the Jews was not unimaginable – it was everything imaginable, exceeding even nature itself by removing all natural death from the Jewish people.

If Thüring mythologized *the Nordic man* in his relation to nature, Darré mythologized the Germanic man of the earth – the peasant – saying, "We National Socialists, who have retrieved the old truth that the blood of the people is the formative element of its culture, see these things with a crystal-clear knowledge. In every period of history the blood of our cities was supplemented by the peasantry and thus the blood of this peasantry repeatedly determined the Germanic content of our city culture."[11] Just as Jewish blood poisons the blood and the soul of the Volk, so does the peasant's blood provide an antidote or vaccine. The peasant's blood thus has a kind of redemptive value: thanks to the essence of nature in the peasants' blood, the Aryan is able to dwell in the city without being corrupted by the urban environment – as long as the Jewish threat has been eliminated.

---

[9] Alfred Rosenberg, "The Earth-Centered Jew Lacks a Soul," in George L. Mosse, ed., *Nazi Culture* (New York, NJ: Grosset & Dunlap, 1966), p. 75.

[10] Bruno Thüring, "Nature Presupposes a Spiritual Disposition," in Mosse, p. 210.

[11] Richard Walther Darré, "The German Peasant Formed German History," in Mosse, p. 148.

In the scheme of Nazi idolatry, then, G-d is not the Creator of humanity; rather, nature is the wellspring of the *Volk, body and soul.* Here G-d does not sanctify nature and all of creation with His commandments; rather, nature sanctifies the *Volk*, not with commandments, however, but, as we have seen, with a certain *Geist* or 'spirit' that is manifest in the passion and the imagination, in the will and the resolve of the people. The Nazis' deification of nature is perfectly in keeping with their elimination of the G-d of Abraham, who sanctifies nature from beyond nature. If, from the Nazi perspective, nature is itself holy, then there is no Holy One who sanctifies it or who might curtail our actions with injunctions such as the prohibition against murder. And so we see the difference between Nazi paganism and Jewish monotheism: whereas the former worships the tree, the latter worships the One who created the tree and who commands us as to how we must handle the tree, as well as how we must treat each other.

Grounded in the embrace of blood and soil, Nazi racism is not about color – it is about idolatry. And the Jews are the primal opponents of idolatry. The Talmud, in fact, goes so far as to say that anyone who repudiates idolatry is a Jew (*Megillah* 13a). Therefore the Nazis were not anti-Semites because they were racists; rather, they were racists because they were anti-Semites. Henri Crétella puts it well: "The adjective *Jewish* does not designate an ethnic group. On the contrary, it signifies that there is no true humanity without being related to divinity—as the Jews have shown us. In other words, it is not blood and soil which properly define us, but rather the possibility of emancipating ourselves from this very blood and soil."[12] Only such an emancipation can enable us to go from serving nature to caring for nature. From the standpoint of Jewish thought, it is not that there is no connection between humanity and nature. No, the teaching is that we are connected to nature through the commandment that comes from the Creator of nature and the Father of humanity. Inasmuch as the Nazis deify nature as the origin of the soul, they defy the Creator and defile His creation.

Because this defilement is part of the assault on the Creator, the extermination of the Jews and Judaism entails the defilement of creation itself. The defilement of creation, however, is not so much a matter of the pollution of nature as it is a question of its deification. And once it is deified, it has its sacrificial offerings: the earth, the air, and the water were saturated not with sulfides, acids, and other pollutants, but with the ashen remains of the Jewish people. The soil once asso-

[12] Henri Crétella, "Self-Destruction," in Alan Milchman and Alan Rosenberg, eds., *Martin Heidegger and the Holocaust* (Atlantic Highlands, NJ: Humanities Press, 1996), p. 159.

ciated with German blood was soaked with Jewish blood and therefore with Jewish essence: ironically, realms were rendered *Judenrein* by being saturated with Jewish remains. The blue Danube ran red with Jewish blood, and the fresh air of the countryside – indeed, of the entire planet – reeked of Jewish ashes. Thus the Nazi environmentalists and nature lovers defiled creation.

With the Nazis' burning of the body of Israel, the earth, entrusted to our care, was covered with the ashes of Israel. Those ashes inhabit the ground from which we harvest our bread. They curl up in the crumbs we put into our mouths. That is where the eclipse of G-d takes place: not in the heavens but on the earth and in the earth, in the very bread upon which we pronounce a blessing and feed to our children. If "the earth is the Lord's and all that is in it" (Psalms 24:1), the earthly trace of the Lord Himself, the Infinite One, is covered with these ashes that breach infinity. Indeed, what we have is what a Nazi declared in Judith Dribben's memoir: with the burning of the body of Israel, "the Jewish G-d is burnt to ashes."[13] As the *Koretzer Rebbe*, a disciple of the *Baal Shem*, taught, "G-d and Torah are one. G-d, Israel, and Torah are one."[14] As Israel is burnt to ashes, so is G-d burnt to ashes; as G-d is burnt to ashes, so is Torah burnt to ashes. And since all of creation is made of Torah, as we are taught (see, for example, *Zohar* I, 5a), creation itself is burnt to ashes.

One task for ecological thinking after Auschwitz is to retrieve the Holy One – to retrieve His Torah – from the Jewish ashes that cover the face of the earth. And it is indeed a *face* that is covered, in Emmanuel Lévinas's sense of the word, even though Lévinas perhaps did not recognize it. Because its face is covered, we no longer suppose that the earth is the Lord's; rather, we think the earth—the Father*land*—is the lord. And so we no longer attend to the commanding Voice that rises up from the earth and all that is in it, revealing itself in the imperative and implicating us for our action or inaction. Instead, we listen to the sounds of wolves and whales, of waterfalls and ocean waves, as we curl up in the complacent illusion that we are 'communing with nature.' Such nature worship is part of the post-Holocaust idolatry that we have inherited from the Nazis. It renders us deaf not only to the Commanding Voice that cries out from the ashen earth, but also to the human outcry that is eclipsed more and more by the 'sounds of na-

---

[13] Judith Dribben, *And some Shall Live* (Jerusalem: Keter Books, 1969), p. 24.

[14] Louis Newman, ed., *The Hasidic Anthology* (New York, NJ: Schocken Books, 1963), p. 147.

ture.' In order to hear the Voice that speaks through nature, as well as the voice of our neighbor, we must attend to the Voice that speaks *from beyond* nature.

HEARING THE VOICE FROM BEYOND NATURE

Inasmuch as the G-d of Abraham, Isaac, and Jacob was the target of the Nazi extermination project, certain teachings from the tradition of the Patriarchs have profound implications for post-Holocaust Jewish thinking about the natural world. The Jewish mystical tradition, for example, teaches that, according to gematria – a method of interpretation based on numerical values of Hebrew words – the word *Elokim* (one of the names of G-d) has the same value as *hateva*, which means 'nature.' Therefore, in the aspect of Elokim, the Creator of heaven and earth (see Genesis 1:1), G-d and nature are of one piece.[15] It is not that nature is G-d, but that nature is, as it were, saturated with the Divine sparks of the Creator's ongoing utterance (see, for example, *Zohar* I, 19b). The assault on G-d as Creator, then, includes an assault on His creation, that is, on nature. With the Nazis, however, as we have seen, the assault on nature is not such much in the mode of pollution as it is in the mode of idolatry, where nature is viewed not as vehicle for the manifestation of the Holy One but rather as the holy itself. Thus made into an idol, nature is rendered mute. Or perhaps better: if nature "speaks," it is the voice of no one –certainly no one who can command us not to murder.

Of course, one must think in Jewish terms in order to determine a connection between *hateva* and *Elokim*, the G-d of Abraham and the Giver of the Torah. As Creator, G-d brings the earth into being through the word, and G-d's word is Torah. Therefore, to hear the Voice that sustains nature – the Voice that speaks from beyond nature – is to hear Torah, which includes the prohibition against murder. And Torah study, as well as the observance of Torah's commandments, can improve our hearing in this regard.

Judaism teaches further that, like nature, the soul is made of G-d's utterance: the soul is made of Torah.[16] The connection between the soul and nature as utterances of G-d lies in the link between *adam* and *adamah*, between the human

[15] See, for example, the commentary of Rabbi Isaiah Horowitz on the first line of Genesis in his *Shnei Luchot HaBrit* (Jerusalem: Kolel Avrekhim HaTotsaat Sefarim "Oz Vehadar," 1993).

[16] See for example, Adin Steinsaltz, *The Sustaining Utterance: Discourses on Chasidic Thought*, ed. and trans. Yehuda Hanegbi (Northvale, NJ: Jason Aronson, 1989), p. 32.

being and the earth. In mystical terms, G-d creates Adam by saying 'Adam.' He creates the earth's soil be saying 'Adamah.' If there is no word for 'the demolition of a man,' as Primo Levi has said,[17] it is because the demolition of a man is the demolition of the word. It is the demolition of the Divine utterance of which *adam* and *adamah* are made. More than a pollution of the earth, the demolition of a man is a desecration of the earth. For the dust of Adam, said the talmudic sage Rabbi Meir, was gathered from all parts of the earth (*Sanhedrin* 38a).

Because the soul is made of Torah, when we damage the soul through a violation of G-d's word, we damage Torah; and since nature is made of Torah, our transgression of Torah does damage to nature. Thus, according to the mystical tradition, every action either elevates or desecrates creation, so that the Nazi evil damages every spark of creation itself. In fact, their idolatrous view of nature is itself part of the assault on nature. From a Jewish standpoint, we refrain from polluting nature not only because of the potential harm to human life but also because the pollution of the earth obstructs the view of the heavens. Which is to say: it blocks the channeling of holiness into the world. Jewish 'environmentalism,' then, is not simply about putting an end to logging or saving the white owl; it is about elevating the material and natural environment to release G-d's hidden holiness into the world. And how is the world, including nature itself, elevated? Through the Divine commandments of Torah. "The mitzvah," Rabbi Adin Steinsaltz states it, "makes an incision into the veil of the hiddenness of G-d."[18] Without the mitzvah, He cannot enter the world. And when He cannot enter the world – when the natural world is itself the object of worship – evil flourishes.

Jewishly understood, then, nature is not an object of adoration; it is a *means* of sanctification. But it is a means that must be employed according to the commandments of Torah. Earth, air, and water harbor a holy fire that can be released when put to proper use, as prescribed in the Torah. The idea is not to be at one with nature – it is to be at one with Torah. We can use wood, for example, to make a Holy Ark for the Torah scrolls, or we can use it to make a weapon. We can use ink to write a Torah scroll, or we can use it to sign a deportation order. We can transform the food we eat into acts of loving kindness or into acts of violence. Whichever action we choose, we either veil the divine sparks hidden in the natural world, or we release them. Either way we create angels, according to

---

[17] Primo Levi, *Survival in Auschwitz: The Nazi Assault on Humanity*, trans. Stuart Woolf (New York, NJ: Simon & Schuster, 1996), p. 26.

[18] Adin Steinsaltz, *The Long Shorter Way: Discourses on Chasidic Thought*, trans. Yehuda Hanegbi (Northvale, NJ: Jason Aronson, 1988), p. 164.

the Talmud, for good or for evil (see *Avot* 4:11; *Chagigah* 41a). As the *Baal Shem Tov* stated it, "of every good deed we do, a good angel is born. Of every bad deed, a bad angel is born."[19] We have no *ethical* relation to nature; animals and trees do not have *rights*; the very notion of right is unintelligible with regard to rocks and reindeer. But we do have a *sacred obligation* – that is, a *commanded relation* – just as we have commandments pertaining to *kashrut* and dress codes, which lie outside the realm of the ethical but nonetheless are essential to drawing holiness into the world. While our treatment of other human beings may be a moral matter, our treatment of nature is a matter of manifesting holiness in the world. What connects each to the other is the Divine commandment to attend to both.

Just as the Torah alone, as the Voice of Hashem, sanctifies human life, so does the Torah alone sanctify the natural world. This uniqueness of Torah belongs to the Oneness of G-d: both instill and sustain all things with the life that is a singularity in being, a breach of being. As we have seen, a basic tenet of the mystical tradition is that everything in the natural world harbors a trace of the Holy One Himself, the G-d of life who is life Himself. Therefore there is no such thing as 'dead' matter. Every atom and every star, every lion and every louse, is a manifestation of a Divine utterance; thus all of nature *speaks*. To be sure, the Zohar teaches that G-d becomes the *Elokim* that is *hateva* in a manifestation of the *Who* (Zohar I, 2a). The *–im* ending can be reversed to make *mi*, meaning 'who,' plus *eleh*, meaning 'all this': as *Elokim*, G-d says to us, 'All this is a *Who*.' And, as a *Who* in all this, He speaks in the imperative. Thus in the Torah and Talmud we have Divine commandments with regard to how we may treat the land and its living creatures. Indeed, the level of soul in the human being called *nefesh* represents a concentration of divinity that is in every living being; therefore at the level of *nefesh* each soul is connected to every other soul, both human and otherwise.[20]

Further, the commandments to give the land a rest (Leviticus 25:4) and to feed our animals before we eat (Talmud Bavli, *Berakhot* 40a), for example, come from the same Origin of Life as the commandment to love our neighbor (Leviticus 19:18). Each is interwoven with the other. Just as the soul suffers what it inflicts in human relation, so does it suffer what it inflicts upon nature. Therefore

---

[19] Quoted in Meyer Levin, ed., *Hassidic Stories* (Tel-Aviv: Greenfield, 1975), p. 47.
[20] See for example, Chayyim Vital, *The Tree of Life*, trans. Donald Wilder Menzi and Zwe Padeh (Northvale, NJ: Jason Aronson, 1999), p. xxxii.

how we treat nature is tied to how we treat one another, precisely because the dimension of height that ordains being deems it so.[21] The environmentalist movements reminiscent of Nazi thinking would eliminate the dimension of height and hierarchy, thus leveling the being of the human being to the status of all natural being. So leveled, the human being has no obligation to nature or to anything else. So leveled, chickens cooped up for slaughter are easily compared to Jews lined up for gas chambers.

Conclusion

What is at stake in a post-Holocaust ecology is a post-Holocaust humanity. The sanctity of both nature and people entails a return to the absolute, Divine commandment to care for both. Only through a renouncement of the idolatrous view of nature that would ascribe to nature the status of god or goddess, where nature would have to care for us, and not we for nature; only though the renouncement of such dangerously absurd categories as 'animal rights,' where we could make certain demands of animals; only through the renouncement of 'environmental ethics,' where we would have to place our treatment of chickens on the same level as our treatment of children – can we determine a commanded responsibility to attend to the care of nature and the outcry of humanity. The postmodern leveling of the human being to the same status as all of nature typifies not only much of the modern environmentalist thinking but also Nazi thinking. "According to our conception of nature," said Nazi bio-racist Paul Brohmer, "man is a link in the chain of living nature just as any other organism."[22] On dozens of environmentalist websites and at conferences on animal rights one can find the same thinking.

A return to that Jewish thinking can enable us to situate G-d, world, and humanity on their proper ground for the well being of all. Neither the modern thought that makes nature into an object of comprehension, nor the postmodern thought that eliminates all absolutes, can determine the ecological care needed not just for the survival but, more importantly, for the sanctification of the planet and its inhabitants. For without the sanctification, there is no reason for the survival.

---

[21] Emmanuel Lévinas, *Collected Philosophical Papers*, trans. Alphonso Lingis (Dordrecht: Martinus Nijhoff, 1987), p. 100.

[22] Paul Brohmer, "The New Biology: Training in Racial Citizenship," in Mosse, p. 87.

In Response to David Patterson

Margaret Brearley

David Patterson's paper, profoundly thoughtful and imbued with clarity of vision and incisive analysis, offers important insights into the Nazi spiritual mission and the linkage between Nazi idolatry, its worship of nature and the Holocaust.

Nazi ideologues deified, muted and ultimately defiled nature in counter-point to their pagan attack on the biblical God: as Patterson notes, "the assault on G-d as Creator [...] includes an assault on His creation, that is, on nature." Moreover, the entire Jewish people, purveyors of God's creative word to the world, was assaulted, initially by demonisation using the language of contagious disease.

Such demonising was itself an assault on God. For in Greek the arrival of a god, the epiphany, was also termed *epidemia*, a "divine 'epidemic' – whose kinship with 'visitation by a disease' is undeniable as it was always the incursion of something overpowering."[1] The Jewish God had overpowered the ancient values and vision of Greek paganism which Nazism aimed to restore; small wonder, then, that God's agents, the Jews, were described in terms of a malign epidemic to be eradicated.

Patterson hints that, in divorcing nature from the biblical God, Nazism envisaged a quite different *epidemia*. He notes that "the Nazis' pagan worship of nature is [...] a spiritualisation of nature into the essence of an invisible god, whose traces are manifest in flora and fauna." I have argued elsewhere that Nazism was attempting in all seriousness to invoke and embody that nature god of terror and ecstasy, Dionysus, whom the Jewish God had ousted.[2] Nietzsche, who had called Dionysus 'the spirit of Germany'. Nietzsche envisaged that his arrival would restore primordial unity among men and between man and nature; man himself would become godlike: "Under the charm of the Dionysian not only is the union between man and man reaffirmed, but nature which has become alienated, hostile, or subjugated, celebrates once more her reconciliation with her lost son, man. Freely, earth proffers her gifts [...] [Man] now feels himself a god, he him-

---

[1] Carl Kerenyi, *Dionysos: Archetypal Image of Indestructible Life* (London: Routledge & Kegan Paul, 1976), p. 139.

[2] Margaret Brearley, "The 'Tempter-God', Evil, and the Shoah" in *Fire in the Ashes: God, Evil, and the Holocaust*, edited by David Patterson and John K.Roth (Seattle, WA: Univerity of Washington Press, 2005), pp. 7-21.

self now walks about enchanted, in ecstasy, like the gods he saw walking in his dreams. He [...] has become a work of art: in these paroxysms of intoxication the artistic power of all nature reveals itself to the highest gratification of the primordial unity."[3] Carl Jung stressed the Dionysian unity of man with animality: "liberation of unbounded instinct, the breaking loose of the unbridled dynamism of animal and divine nature; hence in the Dionysian rout man appears as a satyr, god above and goat below."[4]

It was precisely nature's artistic beauty that led Himmler, chief architect of the Final Solution, to reject hunting: "Nature is so wonderfully beautiful, and every animal has a right to live."[5] Yet the Jewish people had no right to live, since they were viewed as not only racially and morally contaminating but, especially, as embodying the idea of God. For in order to create the sacred space within which the old pagan god could again have pre-eminence, it was necessary to ensure the departure, *apodemia*, of the Jewish God by eradicating the Jews. As Hitler famously noted: "In order to eradicate an idea, it is necessary to eliminate every man, woman and child who bears that idea."

The biblical prophet Job had cried out, "O earth, cover not my blood, and let my cry find no resting-place" (*Job* 16:18). In the Shoah, earth covered blood, bone and ash. Hans Carossa, a National Socialist poet, wrote: "Viel Blut, viel Blut muB in die Erde sinken; nie wird sie sonst den Menschen heimatlich" [Much blood, much blood must sink into the earth; Never will it otherwise be home to humans].[6] As earth drank blood and swallowed ashes, nature was made complicit in mass murder, a Siegfried's *Tarnkappe* (cap of invisibility) to conceal evidence; cliffs were imploded to entomb 33,000 corpses (Babi Yar), fir branches camouflaged barbed-wire paths to gas chambers (Treblinka); trees were planted over ash-pits (Auschwitz). Those who tended nature were co-opted as ideological warriors. State foresters were trained at *NS* bootcamps "to help reform the *Volk* according to the laws of nature as the Nazis saw them realised in the *Dauerwald* (eternal forest)."[7] Göring stated: "Eternal forest and eternal nation are

---

[3] Friedrich Nietzsche, *The Birth of Tragedy* i. Cited Kerenyi, p. 135.

[4] Carl Jung, *Psychological Types* (London: Routledge & Kegan Paul, 1991), p. 138.

[5] Conversation with Felix Kersten, late October 1941; cited in Peter Padfield, *Himmler: Reichsfuehrer-SS* (London: Cassell & Co, 2001; 1st. pubd.1990), p. 351.

[6] Karl-Heinz Schoeps, *Literature and Film in the Third Reich*, trans. Kathleen M. Dell'Orto (Rochester, NJ: Camden House, 2004), p. 175.

[7] Michael Imort, "'Eternal Forest—Eternal *Volk*': The rhetoric and reality of National Socialist forest policy" in Franz-Josef Brueggemeier, Mark Cloc, Thomas Zeller (eds.),

ideas that are indissolubly linked."[8] Auschwitz even fostered the fructification of nature. It contained an SS farm and herb gardens, barns and agricultural laboratories, exemplifying Himmler's chilling linkage of murder with nature's fertility: "It is the curse of greatness that it must step over dead bodies to create new life. Yet we must create new life, we must cleanse the soil or it will never bear fruit."[9] At Auschwitz, human ashes were often used as fertiliser.

Patterson's reflections on the blood and ashes of Jewish victims having polluted nature are profound and have many ramifications. (Ironically, although Nazism tried to limit existing air pollution laws on the principle of 'the collective before the individual', Auschwitz applied in 1943 to the Prussian Institute of Water, Soil and Air Hygiene for a report on the possible environmental impact of its proposed 'heating plant').[10] Jewish blood and ashes both defiled nature and, in one sense, radically reversed the Nazi 'blood and soil' doctrine. Darre had "described how *Volk* and land interacted organically, 'growing into each other' to produce a single entity that was at once natural and national."[11] Instead, land was permeated with ashes of murdered Jews, with signals of God's eclipse by pagan evil.

Paradoxically, there may have been hygienic reasons why earth was made to cover all Jewish remains. For Nazi leaders will have been well aware of (and in some cases complicit in) the genocide in Turkish Asia Minor between 1915 and 1925 of some 3.5 million Armenian, Greek, Assyrian and other Ottoman Christians, may of whom had had typhus or dysentery as a result of their prolonged persecution. Many corpses remained unburied, often rotting in cisterns, wells and

---

How green were the Nazis?: Nature, Environment, and Nation in the Third Reich, (Athens, OH: Ohio University Press, 2005), 43-72, p. 56.

[8] Michael Imort, "'Eternal Forest—Eternal *Volk*", p. 55.

[9] 11.11.1941. Cited Padfield, pp. 324-5.

[10] Frank Uekoetter, "Polycentrism in full Swing: Air Pollution Control in Nazi Germany" in Franz-Josef Brueggemeier, Mark Cloc, Thomas Zeller (eds.), *How green were the Nazis?: Nature, Environment, and Nation in the Third Reich*, (Athens, OH: Ohio University Press, 2005), 101-128, p. 118.

[11] Cited in Mark Bassin, "Blood or Soil? The *Völkisch* Movement, the Nazis, and the Legacy of *Geopolitik*" in Franz-Josef Brüggemeier, Mark Cloc, Thomas Zeller (eds.), *How green were the Nazis?: Nature, Environment, and Nation in the Third Reich*, (Athens, OH: Ohio University Press, 2005), 204-242, p. 214.

rivers. As a result, at least a million Muslims died as a result of contaminated water systems.[12]

Furthermore, Jewish blood and ashes may also may have served deeper symbolic purposes. Exhuming, burning and grinding millions of bones (Maidanek, Belzec, Chelmno, Sobibor, Treblinka) would prevent any Jewish resurrection such as in the prophetic "valley of dry bones" (*Ezekiel* 37, 1-14) and recalled the biblical phrase for ultimate self-humiliation of Jews, "in sackcloth and ashes." In Christian tradition, ashes were used for consecration of sacred spaces (churches). More generally, in some traditional tribal cultures, blood of slain animals must be covered over; the German ethnologist Leo Frobenius commented in *Atlantis* (1921): "it takes a powerful magic to spill blood and not be overtaken by the blood-revenge."[13]

This literal and metaphorical violent disintegration of the Jewish people into innumerable fragments reflects, too, the terrible play of the Dionysian spirit resurrected within Nazism. For essential to ancient secret Dionysian rites was *sparagmos*, dismemberment, often of a living animal (later boiled in milk), as part of 'initiation through mystic sacrifice.' The animal's real suffering, transferred metaphorically to the sacrificer, resulted in his apotheosis. A gold leaf from Greece records the anticipated divine acceptance of the sacrificer: "Welcome, thou who hast suffered such suffering as thou hast never before suffered: From now on thou hast become a god: a kid, thou hast fallen into milk."[14]

Nazism apotheosised Hitler, the supreme Jewish sacrificer, as man-God. Patterson's profound response is to urge reaffirmation of the true God, the God of Abraham, Isaac and Jacob.

---

[12] See Tessa Hofmann's long paper, "Co-victims in Genocide: The Christians during the last Decade of Ottoman rule" (blog.transatlantic-forum.org/index.php/archives/2007/1888/hofmann-genocide-christians/-), p. 69.

[13] Cited in Joseph Campbell, *Primitive Mythology: The Masks of God*, 1st. pubd.1959 (London/New York: Penguin, 1987), p. 29.

[14] Carl Kerenyi, op.cit., pp. 252-253.

In Response to David Patterson

Sarah Pinnock

The Nazi appreciation of nature presents horrific contradictions. David Patterson vividly portrays how the spiritualization of nature was integral to Aryan supremacy and the rationale for killing Jews. The worldview of *Blut und Boden* envisioned the physical and spiritual unity of the German people and homeland, a paradise on earth for the pure race. As I read his essay, I can picture in my mind the opening scenes of Leni Riefenstahl's *The Triumph of the Will* with charming German homes and gardens, like pictures from a tourist brochure. On my first viewing of the film, I remember finding it shocking and sinister how pastoral beauty gives way to militaristic youth marches, massive rallies, and adulation of Hitler. According to Nazi ideology, the purity of this picturesque setting required removing 'unnatural' elements that contaminate the *Third Reich*. The Aryan ideal of nature is juxtaposed with its vicious consequences.

The disturbing impact of this topic is intensified by Patterson's warning about environmentalist thinking today. At intervals throughout his chapter, he asserts that both Nazi ideas and contemporary environmentalist ideas display features that are pagan and idolatrous. As he puts it: "Much of the modern environmentalist thinking devalues human life in the name of a nature that has become the ultimate value; once nature thus becomes an object of worship, anything and anyone can be sacrificed in the name of environmental purity." (119) In his opinion, the belief that the environment is itself holy resembles the naturalistic idolatry of the Nazis. His accusations are addressed broadly, and to support his case, Patterson footnotes a book entitled *Ecofacism* which explores close connections between fascism and ecology in Germany in both the nineteenth and twentieth centuries. Even today, there are right wing groups in post-war Germany that vocally deplore the artificiality of modern urbanism and rationalism and advocate return to nature with romantic and mystical ideals of the German *Volk*. These movements are simultaneously environmentalist and racist, and they obviously repeat Nazi ideas.[1] Patterson's broad warning about environmentalists who consider nature as holy, in contrast with Talmudic thinking, raise questions in my mind about Christian ecological thinking in general.

---

[1] Patterson cites Peter Staudenmaier and Janet Biehl, *Ecofacism: Lessons from the German Experience* (Oakland, CA: AK Press, 1995), pp. 31-41.

Here is a confession. I am sympathetic with visions of organic wholeness between humanity and nature, and I personally relate to nature spirituality. The reason is not loyalty to native soil. My maternal grandparents were Ukrainian farm workers who immigrated to America around 1917, and my paternal parents were British immigrants to Canada. I was born in the US but grew up in Canada, so in a number of ways, I feel lack of belonging to one people or land. My background is evangelical Christian, and emphasis on a personal relationship with Jesus and Bible study hardly seems like a wellspring of pagan tendencies. I conclude that this sympathy stems from my academic studies in philosophy of religion. In particular, my commitment to feminism has drawn me to Christian writers, such as Dorothee Sölle and Rosemary Radford Ruether, who are concerned with ecological issues and nature mysticism. I am aware that Patterson does not direct his criticisms of environmentalist thinking towards feminist authors or towards Christian ecotheology at large. Nevertheless, his chapter raises strong suspicions about the religious valorization of nature correlated with ethical deficiency, which deserve exploration.

As an antidote to Nazi paganism, Patterson develops a Talmudic perspective on nature. He posits a diametrical opposition between Nazi and Jewish thinking, an opposition also endorsed by Nazi thinkers. One key point in his argument is the claim that what was slated for destruction – the Jewish worldview – is urgently needed to defeat the Nazi assault (118). He presents this Jewish or Talmudic worldview as a unity, without considering larger diversity within Judaism. His Talmudic perspective involves recognizing the Commanding Voice of God, the Torah that speaks from beyond nature. It requires the 'dimension of height and hierarchy' to distinguish God and humanity, and humanity from nature. Ethically, Patterson observes that if humanity is on the same level with nature, there is no obligation or reason to object to the killing of human beings like animals for slaughter (126).

In response to Patterson's diametrical opposition between Jewish and pagan environmental outlooks, it is interesting to consider the ecological theology of Roman Catholic feminist theologian Rosemary Radford Ruether. Her work offers an opportunity to explore the compatibility of faith in the God of Abraham and environmentalist thinking where nature is holy. I shall reflect on Ruether's ecotheology in light of Patterson's Talmudic viewpoint, particularly as it contrasts with his emphasis on hierarchy between God, humanity, and nature.

Ruether advocates faith in God that is prophetic, liberationist, feminist, and ecological. In *Gaia and God: An Ecofeminist Theology of Earth Healing*, she considers how to heal the damage wrecked on nature from a Christian perspective. She identifies the covenantal tradition originating in the Hebrew Bible as one healing possibility. Under the covenant, human beings are caretakers of nature, answerable to God. As Christians developed a notion of covenant, what Ruether calls the "concrete eco-justice perspective of Hebrew law" was replaced in early Christianity by a cosmological and spiritualized understanding of the work of Jesus as Messiah.[2] For instance, Christian appropriation of the Jubilee vision of renewal was no longer concerned with practicalities of farming, labor, or economic recompense, but interpreted the Jubilee as a cosmic vision of God's Kingdom achieved in a new heaven and earth. Ruether has a mixed verdict on the Christian covenantal tradition. She indicates the pervasive anthropocentric and patriarchal character of the covenant historically. As an example, she cites the Puritans' understanding of covenant which held that the United States a was land elected and favored by God, while dependent groups such as women, slaves, and Native Americans were subordinated with divine sanction. In her view, the strongest aspect of the Christian covenantal tradition is its prophetic exposure of injustice and responsible stewardship of the earth.

Developing a second healing possibility, Ruether's proposes a cosmological theology and spirituality grounded in the sacramental tradition. She notes that early Christian cosmic theology reflects on pagan religious themes from Hellenistic and Oriental philosophies. For instance, the influence of Platonism led to interpretations of creation as a self-expression of the divine, where God is manifest in the entire cosmos. Ruether acknowledges certain pagan influences on theology, although she points out (in distinction from Platonism) that the incarnation of Jesus Christ involves the Logos taking flesh, which means that the divine permeates bodily nature and thus deifies the bodily.[3] The union of divine and human natures in Jesus departs from both Jewish and pagan thinking. Ruether cites the creation spirituality of Matthew Fox, the evolutionary theology of Telhard de Chardin, and the process theology of John Cobb as examples of a Gaia perspective, where God is relational rather than commanding, immanent rather than transcendent.[4] Such ecotheology distinguishes humans from other life

---

[2] Rosemary Radford Ruether, *Gaia and God: An Ecofeminist Theology of Earth Healing* (New York, NJ: HarperCollins, 1992), p. 215.

[3] Hellenistic and particularly Platonic influences are adapted by the Jewish philosopher Philo as well as many Christian thinkers beginning with the New Testament writers. Rosemary Radford Ruether, *Gaia and God*, p. 231.

[4] Rosemary Radford Ruether, *Gaia and God*, pp. 240-246

forms relatively, but not absolutely. People live in kinship with nature, as all creation is the progeny of Gaia.

Both the covenantal and sacramental traditions – God and Gaia – should coexist in Christian theology and spirituality, according to Ruether. Yet these two perspectives on God are gendered distinctively. God speaks from the mountaintops in masculine tones of commandment, while Gaia speaks from nature not in terms of laws, but inviting humans into communion. At the end of *Gaia and God*, Ruether correlates the covenantal and cosmic views with environmental responses to issues such as pollution and agribusiness. She notes that laws and directives protect nature from human exploitation, while cosmological theology provides an emotional connection to the earth's healing, a sense of unity and belonging. This nature spirituality involves learning to breathe, developing the body-mind connection, participating in nature, creating new liturgies and communities dedicated to global healing. Ecotheology and nature spirituality are necessary to support commitment to protective environmental laws.[5]

It is intriguing to consider whether Ruether's Gaia perspective resonates with aspects of Nazi paganism. She promotes holism between humans and animals, earth spirituality, and it is not an exaggeration to say that she considers nature holy. Moreover, vestiges of paganism inform sacramental theology and modify Hebrew covenantal theology. Her endorsement of intuitive and emotive bonding with Gaia/God is reminiscent of romanticism and mysticism of nature. I do not deny the overwhelming differences between Ruether's view and right wing German environmentalism, but it is worth pondering the ethical effectiveness of her position in comparison with a Talmudic Jewish perspective.

Ruether would disagree with Patterson's claim that listening to the waves and other spiritual communion with nature "renders us deaf not only to the Commanding Voice that cries out from the ashen earth but also to the human outcry that is eclipsed more and more by the 'sounds of nature.'" (122) She would also question the hierarchy of God portrayed by Patterson as the Voice from beyond nature that calls people to responsibility. Rather than share Patterson's emphasis on height and commandment, Ruether understands God through the incarnation, where the Logos of creation is renewed and fulfilled in Christ. Instead of a masculine concept of divine transcendence as separate and beyond, she advocates metaphors of God as Wisdom, mother Gaia, and Liberator of the oppressed. Although she affirms the God of the Hebrew Bible, her sacramental Catholic

[5] Rosemary Radford Ruether, *Gaia and God*, p. 255.

approach opens up to nature spirituality and proposes complementary ways of understanding God in covenant and sacrament.

Ruether's ecofeminist theology offers a third way and a middle ground between the opposites of environmentalist and Talmudic thinking. It is possible that a model of God other than the Commanding Voice can provide critical leverage on Nazi environmental idolatry, and that rejection of vertical hierarchy between God, humanity, and creation does not undermine human obligations to nature and other persons. These issues deserve careful reflection to assess whether post-Holocaust Christian ecological thinking can generate unequivocal opposition to suffering and injustice.

In Reply to My Respondents

David Patterson

I am very grateful to my two respondents, but for different reasons. First, I wish to thank Margaret Brearley for her usual depth of analysis and insight and for helping me to make my case. Her points about the relation between *epidemia* and *apodemia* are especially helpful in demonstrating the Nazis' ideological necessity to exterminate the God of Abraham through the extermination of the Jews. I think she is quite right in noting that the pagan resurrection of the Dionysian "god of terror and ecstasy" is reflected in National Socialism, a move in which "[Man] now feels himself a god" (80). This creation of the mangod through a pagan union with nature underlies what Brearley demonstrates to be a link between murder and nature's fertility (81), a link that is itself linked to National Socialist environmentalism.

I am also grateful to Sarah Pinnock for the questions she raises and for an opportunity to offer some clarification. With regard to the contrasts she outlines between my Torah-Talmud-based Judaism and Rosemary Radford Ruether's views, some of the contrast lies in a general difference between Judaism and Christianity, and Pinnock is aware of this. The Christian doctrine of the Godman manifest in the Incarnation, for example, may well lend itself to the merging of nature and spirit that one finds in Ruether's ecofeminist theology. The language of communion with nature also has suggestive overtones with regard to the Eucharist.

Other contrasting points, however, lie in some common misconceptions concerning Jewish teaching. For example, the height and hierarchy that I espouse does not preclude the kind of divine immanence that would make the act of eating, for example, a sacred, even eucharistic act. According to Jewish teaching, the biblical verse "man lives not by bread alone, but by every utterance from the mouth of God" (Deuteronomy 8:3) does not mean that we have a physical aspect distinct from a spiritual aspect, as it is often misunderstood. No, it means that the bread is not *merely* bread—it is also a divine utterance. Placing it into our mouth, we ingest a divine Word, which commands us to transform the bread into deeds of loving kindness. Because all that exists derives its existence from a divine utterance, God is indeed "manifest in the entire cosmos," as Ruether says (85). Therefore to contrast ecofeminist theology and Judaism by saying, "God is…

immanent rather than transcendent" (85) is to misunderstand Jewish teaching. According to Jewish teaching, the physical is also metaphysical; the material is also spiritual.

There is a similar misunderstanding in the contrast between a "relational" God and a "commanding" God (85). In Judaism the act of creation is understood as a movement into relation, as indicated by the word *bara*, the word meaning "created" (Genesis 1:1). For *bara* is a cognate of *brit*, which is "covenant." As an act of entering into a relation, the act of creation entails the emergence of the You in all of creation. Instead of reading *Bereshit bara Elohim et ha-...* as "In the beginning God created the...," the Zohar reads it as "In the beginning God created the *alef*, *tav*, *hey* of *atah*: You" (*Zohar* I, 15b). This You-saying on the part of the Creator commands us to enter into a relation with Him by entering into a relation with one another. It is a You-saying steeped on love, and through love the Creator shows us how to connect with Him. And the connection – the loving relation – lies in the commandment.

Indeed, the commandment to "love your neighbor as yourself" (Leviticus 19:18) is a commandment to be who we are, for who we are lies in just such a relation. "As yourself" is *k'mokha*, or "that is what you are." It does doe mean "I know how much you love yourself, and that is how much you should love your neighbor." And the word for "you shall love," *v'ahavta*, is not a commandment to have a certain feeling but to enter into a certain relation by performing a certain action, the action of giving, as the verb implies: the root of the verb to "love" is *hav*, which means to "give." The commandment is the key to the relation, both to neighbor and to nature. Because it is better to be in the relation than not to be in it, we have a hierarchy defined by the commandment.

Being commanded by the Creator, then, is essential to being in a relation with the Creator. This becomes clear when we recall that word for "commandment" is *mitzvah*, whose root is *tzavta*, a word meaning "connection." As a parent connects with a child out of love, so God connects with the human being out of love. One way a parent expresses this loving relation is through commandment: "Thou shalt not run into the street without looking for cars." In Judaism this love expressed by a commanding God is often represented as a maternal love. In the Zohar, for example, it is written: "First came *Ehyeh* (I Shall Be), the dark womb of all. Then *Asher Ehyeh* (That I Am), indicating the readiness of the Mother to beget all" (*Zohar* III, 65b). The "I Shall Be" posits the yet-to-be that is the horizon of meaning; the "That I Am" is the manifestation of meaning along that horizon: begetting all, the Mother begets meaning, both in the natural and in the spiritual realms. And She begets love: the word for "womb" is *rechem*, which is a cognate of *racham*, meaning to "love" or to "have compassion."

Speaking of the Creator in masculine terms, then, can be misleading, and Jewish teaching is very much aware of that. Begetting all, says the Zohar, the Supernal Mother begets all of humanity: "The [Supernal] Mother said: 'Let us make man in our image'" (*Zohar* I, 22b). This teaching is based on letter *beit*, the first letter of the Torah, which is shaped like a womb: the creation of heaven and earth is an act of birth. As the Creator of heaven *and* earth, the Supernal Mother is not reducible to the Earth Mother or Gaia; from a Jewish standpoint, such a reduction is a pagan move. It is a move that would identify nature as holy, something which Ruether does, as Pinnock points out (85). But to say that nature itself is holy, rather than sanctified by the Holy One, runs contrary to the very meaning of the word *kodosh*, the word for "holy." It does not designate something really precious or very good. No, the word means "separate" or "distinct" from all other things. *Kadosh* does not refer to one special thing in the ontological landscape of things; rather, it designates what lies outside of all ontological categories and therefore what imparts meaning to ontological reality. It is precisely what is *otherwise than being*, at once immanent and transcendent. How we treat nature matters—nature itself matters—because the One who is both beyond and within nature has made it so.

Finally, Pinnock implies a contrast between tradition-based Judaism and Ruether's position by saying, "Instead of a masculine concept of divine transcendence as separate and beyond, she advocates metaphors of God as Wisdom, mother Gaia, and Liberator of the oppressed" (86). I have tried to make it clear that in Judaism divine transcendence is not necessarily masculine, nor does it preclude divine immanence. Indeed, what is manifest in immanence is precisely the transcendent. In Judaism the teaching that God creates through Wisdom (*Chokhmah*, a feminine noun) is as familiar as the teaching that God creates through Torah. I see no point of contrast here. Nor do I see a contrast in viewing God as the "Liberator of the oppressed," since among the ways we refer to God is to say He is the *Avi Yetomim*, the "Father of the Fatherless," and the *Dayan Almanot*, the "Protector of Widows" (as in Psalms 68:6). God attends to the care of the widow, the orphan, and the stranger (see Deuteronomy 10:18) by commanding us to do so, just as He commands us to attend to the care of the natural world. God liberates us from the horrifying neutrality of *mere* nature through His Commanding and Caring Voice.

The identification of God with the mother Gaia, however, is a point of contrast. When divorced from the other stipulations that Ruether makes, such an identification might play into the hands of something like Nazi paganism. Pinnock's suspicion that this identification may be consistent with Christian theology suggests that there might be something pagan about Christianity; but that is a topic for another argument. What history has demonstrated, I think, is this: the privileging of passion and emotion over the Commanding Voice—a Voice that commands us to feed the hungry and care for the environment whether we *feel* like it or not—may lead us to going wherever our passion may take us. From there it is a short path to the resurrection of the Dionysian "god of terror and ecstasy."

The Bible in the Ecological Debate:

Obstacle or Guide?

Didier Pollefeyt

INTRODUCTION: BIBLE AND ECOLOGICAL CRISIS

Ever since the beginning of ecological thinking, the Jewish and Christian religious traditions have been identified as one of the most important causes for today's environmental crisis.[1] Biblical texts, the argument goes, would be responsible for exalting man to the position of the anthropocentric pinnacle of nature. Being the only creature made in God's likeness, man's mission is to 'have dominion' over nature and to 'subdue' it (Gen 1:26-28).[2] By such phrases the Bible seems to suggest that nature is merely the object of man's capriciousness and self-glorification (*cf.* instrumentalist anthropocentrism, as denounced by Gaia's film about battery cages for chickens). Or, as it was well put by a student, the proposition that (only) 'man is made after God's (imageless) image' is a pretentious Jewish-Christian statement.

In the present chapter, we intend to take this criticism seriously by means of a philosophical in-depth reading of the Bible. We are inspired in this by the thought of the Jewish philosopher Catherine Chalier,[3] a former pupil of Emmanuel Levinas. In doing so we will try to uncover a 'forgotten' dimension of the Jewish-Christian tradition, namely the *connectedness* of the Biblical concept of God with the whole of creation (Jer 33:25).

Chalier says that the demise of the 'ecological' dimension within the Jewish tradition is connected to the historical experiences of exile of the Jewish people, experiences that brought with them an alienation from their country for the duration of centuries and a forced dissociation from nature and its rhythms. To the extent that the Jewish people were 'tolerated' by foreign societies, they were also systematically denied immediate and intimate contact with nature. Moreover, this enforced

---

[1] For an apt example, see Lynn White's remarkable and provocative article, "The Historical Roots of our Ecological Crisis," in *Science* 155 (1967), 1203-1207.
[2] Johan De Tavernier, "Ecologie en ethiek," in *Collationes* 23 (1993), 393-418, p. 403.
[3] Cathérine Chalier, *L'alliance avec la nature* (La nuit surveillée) (Paris : Cerf, 1989), p. 211.

reticence towards nature has always had an apologetic function in Judaism, as it allowed people to distance themselves from pagan idolatry and the deification of natural and cosmic forces. On the downside, however, this defensive attitude led to the loss of the idea that the path to the secret runs *through* nature as the work of God. The great Jewish liturgical festivals, which commemorate historical events, were celebrated without reference to the moments in nature's cycle with which they coincide and which they factually celebrate. Nature only was a comfortless desolation that surrenders man to alien, depersonalising forces of being (*il y a*) (Levinas). God is totally 'other', completely different from the world (*autrement qu'être*), hidden in a total transcendence, which has no reference point whatsoever in the 'good' creation.
However, man's forlorn state, living in a cold and indifferent world is neither the first nor the last that is said about man's relationship to nature in the Bible. On the contrary, this chapter will show that, in the Bible, the covenant between God and man takes shape within the heart of creation itself, that nature has been touched by the same creative breath from which man has sprung, and that all things have a common goal.

THE FACE OF NATURE?

When in 1991 we visited Catherine Chalier's teacher, Emmanuel Levinas, in Paris in the company of a group of students, one of my students asked him whether 'an animal has a face'. Levinas was visibly surprised by the question. In his thinking nature is understood chiefly as *il y a*, some sort of a formless and impersonal swarming, as 'being without a face', as that which depersonalises. This view of nature can be elucidated by a quick look at the *Sitz im Leben* from which Levinas' thought has grown.[4] The notion of *il y a* was first developed in Levinas' book *De l'existence à l'existant*, which was written during his internment in a *Wehrmacht* camp in Hannover in 1939. In the camp, Levinas and some fellow-Jews were assigned to a special command that had to carry out heavy duty labour in the woods nearby. During his days in the labour camp, Levinas went through a grim existential *struggle for life* against the depersonalising forces of nature. His notion of *il y a* can thus be seen as a philosophical translation of this experience. It is then also

---

[4] See our contribution Didier Pollefeyt, "The Trauma of the Holocaust as a Central Challenge of Levinas' Ethical and Theological Thought," in *Remembering for the Future II* (Oxford/New York, NJ: Pergamon Press, 1994).

quite evident that Levinas did not became a lover of nature and rather turned to the city in later life. For Levinas, philosophy does not start from the miracle of nature (as it does for Heidegger, who speaks of the 'lights of being'), but from the trauma of evil.[5] He holds that God reveals Himself in the vulnerable face of the other, which can take down every fragmentation, and not in the merciless, unpredictable forces of nature that harm man's vulnerability. This may explain the fact that Levinas has not developed his thought on the level of ecology.

Unlike her mentor, Chalier does take up the challenge of ecological thinking from the perspective of Levinas' thought. Where Levinas speaks of God revealing Himself in the face of the other, Chalier speaks of God revealing Himself in the *traces* He has left in nature. With her notion of 'the trace of God', Chalier combats_two one-sided views on the relationship between God and nature: the pagan identification of God with nature on one hand, and the modern day desacralisation of nature on the other.

According to a later interpretation of the creation story, God creates being out of nothing (*creatio ex nihilo*) through the Word. The concept of creation implies God's transcendence over the world. Through the creation God calls into existence something other than Himself. As a matter of fact, the Bible thus articulates a sharp criticism of pagan practices that deify nature. The Biblical God is not the diffuse, supportive ground for Being that exercises a 'fascinating' and 'frightening' attraction over man which is so great that man wants to participate in this ground and wishes to dissolve in it. Such a *Gott mit uns* would no longer be a critical, Biblical 'opposite', but a numinous power that on the one hand stirs up the human being to a blind and irrational enthusiasm, but on the other hand spreads an arbitrary terror that dissolves the basis for any kind of personal response-ability.

The Jewish tradition holds that God's glory exists precisely in the fact that He has placed someone in his creation who can seek Him out in his separation and who is *in the ability* of being responsive to Him (though not obliged to do so). Holding on to the absolute transcendence of the Creator implies the possibility of atheism. Man can experience the irreversible separation between God and the world as an enormous absence. Human beings are in danger of being overwhelmed by the inhuman neutrality of a silent and obscure cosmos.

---

[5] Didier Pollefeyt & Luc Anckaert, "Tussen trauma en verwondering. Rosenzweig, Levinas en Fackenheim," in Bart Raymaekers, ed., *Gehelen en fragmenten. De vele gezichten van de filosofie* (Louvain: University Press Leuven, 1993), 159-164.

The distance between God and the world, however, is not absolute for the Bible. Nature is not merely the atheist, threatening *il y a* that has to be controlled. The entire cosmos contains Traces of God's creative actions. Man is called to uncover and unravel the Traces that God has left in his Creation, and to bring new life to their meaning. Yet, this presupposes a hermeneutical attitude towards nature on the part of the believer.

TOWARDS A HERMENEUTICS OF NATURE

The tradition often attributes the Jewish forgetfulness towards nature to the rabbinic passion for the study of the Scripture. The Jewish exegete searches after the power of the text (hence not a literal, fundamentalist interpretation of the Bible). He closely examines the verses in hopes of finding (previously) hidden faces. Because of this, however, the rabbi not infrequently progressively forgets the necessity to contemplate that other great riddle: creation.

Although nature is not divine, it does testify of God. It can just as well be seen as a great 'Book' (a text) which has to be read and interpreted to (get to) know He who has left His Signature in the whole of Creation. The Creator of nature and the Giver of the Torah are thus one and the same God. Especially Chassidism has taken this other route towards finding God's love, rather than studying the Scripture. Chassidism has returned to careful listening to the earth and the heavens which speak of the Beauty of the Eternal on earth. Its followers are taught that nature is the place where Abraham, Isaac, and Jacob have found revelation.

The fact that God can also be found in that other book (nature), in no way means the end of the relevance of the Scripture. On the contrary, the Scripture itself tells us to look to nature as a work wherein God has left his Traces. Without the Scripture, man would never be able to listen to Creation as the site of God's revelation. When we would read nature *in opposition to* the Scripture, we would never be able to find that of what nature is the sign. Then, the temptation of paganism, wherein nature itself is exalted to being an ultimate, divine reality (*cf.* the theology of Rubenstein), would arise.

The interdependence of reading nature and reading the Scripture even stretches beyond this. For nature can not be read differently from the words in the Scripture. Contemplating nature as a 'riddle', which means that nature is thought of as receptive to a hermeneutics (or interpretation), is principally impossible for some one who has not learned to read the Scripture. 'Hermeneutics' can be placed in opposition to 'dogma' here. A dogma is posited without riposte, as if the utterance would once and for all be fixed in the unchanging character of the letter. The Jewish tradition, conversely, holds that the Torah has seventy faces. The Torah is as it were waiting for every (irreplaceable) generation of readers. The Zohar, the collection of influential mystical comments on the five books of Moses, calls for heavenly joy for every new interpretation of the Scripture. Because it is not the literal meaning that matters, every new reader is important. The Scripture needs to be taken up time and time again as a pathway to its secret, the transcendence, which as it were must be begged to the surface. Without such an exegesis as hermeneutics, the Scripture would be meaningless for us, like a flame without a wick, slowly dying out.

The old imperative to search for Traces of God in the humility of the verses and the letters of the Bible, to search for the part of the secret which it still harbours, now *mutatis mutandis* also applies to the Traces that God has left in the 'clods of the earth' (Job 38:38) and the 'rocky crags' (Job 39:28). For, like the Scripture, nature presents its riddle to us as a language that asks to be interpreted. Truth sprouts from nature, similar to the way it reveals itself to a student of the Scripture.

In Chalier's thought, the Scripture thus is the necessary mediation between man and nature. Without the Scripture, man runs the risk of contenting himself with the immanent beauty of nature. Pantheism is not far off in such a case. The study of the Scripture, on the other hand, teaches man to orient himself to the Infinite that is revealed in and through the cosmos, but which is not the cosmos itself. The Scripture teaches man to see beyond that his own (literal) horizon. No matter how incredibly small the Scripture may be in comparison with the overwhelming dimensions of nature, it still offers the perspective through which nature *can* reveal itself as the Word of God.

MAN: LORD AND MASTER OVER NATURE?

A hermeneutical openness to nature as God's creation such as the one described above is not quite as evident as it may seem. A good example of the need for an apt attitude to see Creation as God's revelation is the story of Job. After a long period of remaining terribly silent to Job's protest, God suddenly decides to reply to Job with an inventory of the richness of His Creation. Job thus does not get the answer he was hoping for: a theoretical explanation or some words of comfort are not on God's mind. God just presents Job with His Creation as if His answer to Job's misery lies there. He speaks to Job about the coming to being of Creation: 'Where were you when I laid the foundation of the earth?' (Job 38:4). The beauty of the

Creator which shines through in all things, great and small, had been unnoticed by Job. Seeing this beauty with own eyes, as he does from that moment onwards, was something that had to be taught to him. His misery was an obstacle in his understanding of the language of the Creation. The miserable are condemned to live within the confined limits of their own ill bodies. They are hounded by their own vulnerability, which turns every contact with the world into inhuman torture. The hermeneutical paralysis of suffering men is a consequence of their inability to safeguard some space in themselves where the other can be received. Job's immense sufferings deprived him of the ability to look to nature through a different frame of mind, and not merely see it as something that just pursues its own course, totally indifferent to his misery, almost mocking him. How great is not the temptation to see nature merely as an eternal and vain frivolity (Eccl 1:5-6)?

The receptivity for that specific, 'appealing' opening in nature, that unique marking in nature which orients man to an *au-delà*, is hard to experience for someone (miserable) who is strongly bound to his own being. When man is absorbed in his interest for his own being (Levinas), he will not be able to see nature as the work of God wherein He has left a Trace. God only shows Himself to those who are receptive to His Traces. In extreme circumstances, this receptive attitude can best be described as sanctity. A saint is someone who always leaves space in himself for the beautiful, even when he is filled with and surrounded by nothing but death and destruction. Etty Hillesum describes the intensity of being touched by a blooming jasmine that was reaching up to the blue sky in the mud of the Nazi camp in Westerbork. In a place where all is lost and abandoned, Hillesum learns to listen to nature, as if the sense for the other, hidden in nature, can only be found in places where all human and natural sumptuousness has been discarded, where man is thrown back upon his lowest degree of being.[6] Chalier calls experiences like that of Hillesum 'desert experiences'. In the desert, man is stripped of everything, initiated in the humility of being deprived of every form of possession and almost forced into an extreme listening to the meaning that comes out of the paucity of things. The Hebrew language holds an immemorial connection between 'the one who speaks' (*medaber*) and 'the desert' (*midbar*). In the desert of Sinaï the Jewish people, still burdened by their suffering as slaves and with the hardship of their passage on their minds, received the Torah. It seems as if they had to go through the experience of the great prohibition to appropriate things before they could enter the

---

[6] Etty Hillesum, *De nagelaten geschriften van Etty Hillesum 1941-1943*, Klaas.A.D. Smelik (Amsterdam: Uitgeverij Balans, 1991, third revised edition).

Promised Land, a land of 'milk and honey' (Ex 3:8), 'a land of wheat and barley, of vines and fig trees and pomegranates, a land of olive trees and honey' (Deut 8:8).

In the desert, man discovers that he is not always condemned to turn back upon himself but that he can open himself to the other which pulls him away from himself and which frees him of himself. The contemplation of nature can dis-engage man from his own individuality and lift him above his own interest. This way the infinite can invade his existence and order him to give up the things that normally keep him busy (health, money, and life) to the benefit of a destination that transcends the narrow limits of his own interest.

Precisely this relation to nature has become problematic today. In modernity, man has set himself up as the *maître et possesseur* of nature (Descartes). The physical world has been reduced to its mathematical dimensions and has been brought to silence. Although the world is still an enormous book 'written in a mathematical language' (Galilei), it no longer speaks to the human heart. The modern world no longer participates in an attentive hermeneutics of nature, but forces nature only to answer the self-interested questions that man asks it. It is not the 'exegesis' which is central in our understanding of nature today, but the 'genesis' of nature. Modern Bible exegesis is often limited to a study of the Bible's 'genesis', which is a study of the social, historical and literary background of the Bible texts, as if those collected texts are merely a worn-out fossil that no longer sets one to thinking.[7] Similarly, modern science mostly focuses on the laws of physics without believing in a 'talmudic' reading of nature, which is a reading that brings to light the new, current, and unspoiled power that speaks from nature. The strict, mathematical approach of science thus has become the only legitimate approach for studying nature (and the Scripture). This, in turn, leads to a state of affairs wherein the proud theoretical study of the genesis of natural phenomena replaces exegesis' humble singular search for meaning. Man's inexhaustible urge for control reduces and substitutes the speaking power of nature. This brings along the deep existential fear that sometimes takes a hold of people when they discover that nature's riddle will in the end always be undecipherable. At such moments, science leaves us with many uncertainties about the place and the meaning of our presence on earth. A saying by Pascal is exceptionally paradigmatic for this feeling: "*le silence eternal de ces*

---

[7] Roger Burggraeve, *De bijbel geeft te denken: schepping, milieu, lijden, roeping, Gods passie en de ander, vergeving, bevrijding van de ethiek, in gesprek met Levinas* (Louvain: Acco, 1991), chapter 1.

espaces inifinis m'effraie (the eternal silence of those infinite spaces frighten me)."[8] With this phrase, Pascal expresses the panic of modern man who feels himself to be radically alien in a universe that has been constructed by the measuring and calculating mind, a universe in which order has replaced interpretation.

NATURE AS A MEETING PLACE WITH THE OTHER

As such, the modern, totalising subject of the *Aufklärung* is the most important obstacle for a hermeneutics of nature as a work of God. The ideal of scientific objectivity makes modern man lose its sense of humbleness: he is no longer capable to receive within himself that which goes beyond his self-interested concepts and theories. As lord and master over nature, modern man has lost every openness for a meaning and sense which gives itself in the form of an infiniteness, and which at the same time also retreats itself in its giving as the humbleness of a Trace. By confining reality in a network of concepts and theories, a hermeneutical interaction with that which will always throw up resistance as alternity and exteriority has been totally lost. Instead of astonishment for things that will always resist its reductions as 'the other', science has developed a deep aversion for the riddle of nature, a riddle which nonetheless holds a secret that should encourage people to a different kind of hermeneutical thinking.

Watching and listening to nature with modesty and dis-interestedness, without wanting to immediately claim and posses it, is in other words a prerequisite for the welcoming of the infinite in the finite. The meaning of nature as a Trace of God's creation will only present itself when man is able to reserve a space within himself for the other as other. This other does not force itself on man, but gives itself in the discretion of a presence that always retreats at the moment that it is in danger of being trapped by the concept. Thus, it is not so much a matter of apprehension of the other, but rather a matter of receiving the other, averse to any violent conceptualisation, and up to the point of shuddering for the fragility of this other. For Chalier, this disinterestedness contains, precisely by its opening up and redirection of our own needs and interests, the key to a new, ethic relationship with nature. When man is called to 'subdue' nature, this does not mean that he is called to abuse it, shamelessly exploit or reduce it to whatever profits one can get out of it. It is the submission of that which embodies the Trace of creation, of that which does not coincide with being human and which is never to be reduced to the human. Man

[8] Blaise Pascal, *Pensées* (Paris, Flammarion, 1973), nr. 91.

has to abandon a purely reductionist view of nature. When he discovers the Traces of creation in the things he controls and cultivates, he will also become aware of the fact that he is not *chez soi* in this world, that he fundamentally is and remains a stranger.

The ethical encounter with nature thus presupposes the ethical excellence of the subject, to the extent that it is capable of orienting itself towards the other, without continually returning to itself. Only such an ethical subjectivity can be witness of nature as a work of God. Ethics opens up the immanence of the natural order and the human control over it by opening itself for an *au-delà* that has been left in it as a divine Trace. Without ethics, in other words, the riddle of nature stays lost, distancing itself in nature's violent indifference. The meaning of nature can thus not be deduced from some rational or technical analysis, similar to the way in which natural theology searched for the great motor of the universe. The riddle's meaning is radically dependent on man's readiness to reshape itself to an openness, to a meeting place wherein the other can live. Only the disinterested humbleness of the mind and the heart opens up the room that is necessary for a contemplation of nature as a work of God. Without the ethical subjectivity the riddle of nature finally withdraws behind an unreachable horizon.

THE MIRACLE OF NATURE?

Of course, the question remains whether 'after Auschwitz' such a view of nature is not dreadfully naive. In the concentration camps nature has not only shown itself as utterly indifferent to man's fate (the flowers were equally beautiful in Auschwitz), but also as a supplementary source of suffering (cold, hunger). How can one in the century of Auschwitz still speak of the divine 'miracle' of nature? Still, many in the concentration camps were able to retain their ability to, with astounding mental clarity, receive the birth of every new day as a pathway to the other and a gift from God. Many Jews in Auschwitz also continued their prayers and the celebration of liturgical holidays, wherein, even more than before, the connection with nature's cycles played an important role. They again looked to the sun and the stars as God presented them at the beginning of creation: 'for signs and for seasons and for days and years' (Gen 1:14).

That is why Chalier still dares to speak of the 'miracle' of creation in the face of Auschwitz. Thanks to people like Hillesum, who *in* Auschwitz have seen nature as a Trace of God, we are still able to perceive nature as a work of God *after* Auschwitz. The experiences of Holocaust victims do not only make this possible, but they also categorically call upon us to not condemn man to a cosmic solitude because of Auschwitz (Fackenheim).[9] Seeing nature as a 'faceless abyss' or a 'cannibal Earth-Mother that can only be appeased and satisfied by human offerings', as the Jewish 'Holocaust theologist' Richard Rubenstein does,[10] would amount to giving a posthumous victory to Hitler. The sacralisation of the immanent forces of nature and the *Wille zur Macht* were central concepts in Hitler's *Weltanschauung* (*cf.* the classes: Nazism as an *avant-la-lettre* ecological movement).

At this point, we are able to formulate a critique of the way in which the 'miracle' is usually understood, namely as a random abolition of the natural order. Such an understanding of the miracle, however, reduces God's diligence to 'what is good for my own being'. Such a God becomes a *Gott mit uns*. Yet, God's created nature obeys a regularity of laws that cannot be altered by the Creator. The medieval Jewish philosopher Maimonides (1135 – 1204) already said that the miracle is a possibility of nature, and not a consequence of an abolition of nature's laws, laws that actually keep nature in existence. When God intervenes in natural processes, he does not damage the laws of physics according to Maimonides, but rather uses them to their optimal effect to His wishes. For Chalier, the miracle does not happen a lot in the noise of great events, but rather discretely in the heart of daily events. We have to leave the prevailing idea of the miracle as an adaptation of the other to the will of the same. Exactly the opposite happens in the miracle: the invasion of the order of the same by the intrusion of the other. The miracle does not allow for a human explanation, as it immediately exceeds the limits of the understanding individual as the entrance of the other in the same.

The pre-eminent miracle is creation itself, not just as a singular divine act in a distant past, but as a wondrous event that keeps repeating itself in the present. For believers, God continues to create reality at every moment. The Jewish sabbatical year is a good illustration of this. When Jews stop working, sowing and harvesting for an entire year every seven years on the basis of a

---

[9] Emile Fackenheim, "Damit die Erde menschlich bleibt: gemeinsame Verantwortung von Juden und Christen für die Zukunft," in G.B. Ginzel, ed., *Auschwitz als Herausforderung für Jüden und Christen* (Heidelberg: Schneider, 1980), pp. 86-112. Also see our study D. Pollefeyt, *De Holocaust: de verhouding tussen theologie en ethiek op een keerpunt? Confrontatie met de joodse visie van Emil L. Fackenheim* (Unpublished master dissertation Theology, Louvain, 1991), p. 313, pp. 296-300.

[10] Richard Rubenstein, *De God van de joden na Auschwitz*. Translated from English by P. Telder (Utrecht: Ambo, 1968). Also see our study Didier Pollefeyt, *De Holocaust: het einde van theologie en ethiek? Confrontatie met de joodse visie van Richard L. Rubenstein* (Unpublished master dissertation Religious Studies: Louvain, 1988), p. 35-40.

commandment of absolute rest for man and animal (Lev 25:2-7), they do not only express a complete distancing from the unlimited dominion over nature, but they also come very close to the idea of the continuous recreation of nature. Leaving the earth to itself, allowing it to rest completely, reminds man of the fact that he does not fully posses the earth, but also of the internal impetus that is at work in creation and which should be respected. Creation is not only a divine gift (Deut 21:1) which we can treat according to our own discretion, it is also animated by an unstoppable force that by definition escapes human omnipotence. The sabbatical year reminds us of the fact that 'creating' is not a singular past event, but that the creation produces itself constantly, again and again every moment. All things are created out of nothing all the time. Without the continuously animating breath of God that constitutes the very inner of every being, things would relapse into nothingness. Charlier calls the idea of a continuous creation the foundation of God's own Name: 'I am who am' (אהיה אשר אהיה)(Ex 3:14). God's promise to Moses is not only a pledge to never abandon Israel in the course of history, but it is also an expression of loyalty to the durability of all nature's life. God reveals a fundamental secret to Moses: he teaches him to recognize the infinite, divine life in the finite.

THE MESSIANIC CREATIVE ASSIGNMENT OF MAN

In the book Ezekiel, we find the idea that the messianic peace concerns both nature and the human community (Ezek 34:24-29). There is no hope of peace at the end of times when relationships between humans are good while the violation of nature continues. Reconciliation between manhood and nature is also necessary.

For the prophet Isaiah, it is clear that the totality of creation awaits the end of times and the exile. As man has dragged nature along in his fall, nature will also participate in man's rebirth.

> The wolf shall live with the lamb, the leopard shall lie down with the kid, the calf and the lion and the fatling together, and a little child shall lead them. The cow and the bear shall graze, their young shall lie down together; and the lion shall eat straw like the ox. The nursing child shall play over the hole of the asp, and the weaned child shall put its hand on the adder's den. They will not hurt or destroy on all my holy mountain; for the earth will be full of the knowledge of the Lord as the waters cover the sea. (Isaiah 11:6-9)

The radical renewal of creation thus does not only concern man. For the Bible, it is no use speaking of salvation as long as the desert keeps its dryness, the fields their barrenness and the animals their cruelty. Moreover, the hope of salvation goes a lot further than the mere restoration of the order that was destroyed by sin and suffering. In the prophetic texts, a new reality is announced: a new heaven and a new earth.

THE DIFFERENCE BETWEEN MAN AND ANIMAL

Although not only man, but the entirety of creation has been taken up in the covenant with God, it is only man who has been called on to hold the responsibility for nature's survival. An animal only follows its instincts and does not experience a desire to transcend its own nature (*cf.* the anthropology that we developed in the first session). It is not aware of the golden thread of the inner that links man to the transcendent. It does not have that intimate and fragile place where the transcendent can be received. The animal's drama is the radical dissociation of the inner and the outer. The animal is darkened by the power of its own impulses. It is forced to find satisfaction in the outer world, and at the same time this misty impulsivity obstructs any possibility to make space to receive meaning. Abraham's departure from Ur is a powerful symbol of man letting go of the animal ties to nature, and his going on, from horizon to horizon, to find the meaning of existence in a dialogue with the Other (and this without returning to oneself, in contrast to Greek philosophy). Only man is capable of such a demanding interiority which enables him to receive the word of the Other.

It is *this* human being who is summoned to rule and have dominion over all creatures. Exactly in this responsibility for the entire universe lies man's calling and his unique being-image of an imageless God. Man is the only creature that can distance itself from itself, even if it is but during one moment of his life. Only man can 'disinterest' himself (*cf.* man as an ethic subject, the animal as an ethic object). I absolutely do not deny that self-interest is (or can be) healthy (*cf.* the optimistic anthropology: man is fundamentally oriented towards the good, also on the level of his own corporality), but I do want to ask whether man is not a murderer when he is only healthy.

Evidently, the necessity to appeal on nature's resources for man's needs will remain a fact of life, even if nature is transformed by man. It goes without saying that modern science is an improvement over premodern man's fear for the numinous, unpredictable forces of nature. Yet still, this known fact does not necessarily have to lead to the exploitation of nature for blind profits or man's tyrannical urges. Cultivating the earth, watching over the plants, descending into the heart of matter to distil life energy from it and even eating animal flesh to alleviate one's hunger, do not inevitably imply the destructive exploitation of natural resources and extorting animal life for commercial purposes. The first attitude holds on to the sense for the other. The second attitude cancels out this reference to alterity and complacently settles for a purely reductionist attitude wherein in the end only the interest for one's own being is the norm. The first position is enlightened by a concern for the good that gives meaning to human actions (this can for instance take form in an ordering of non-human life on the basis of a 'pathocentrism'). The second approach reduces itself to a functional rationality that contents itself with a limited concern for one's own being only.

Plea for an Ethically Qualified Anthropocentrism

This chapter has shown how the recovery (*tikkun*) of the world, broken by man's and nature's suffering, is inseparably connected with altered thinking. Only human beings that are oriented by alterity can maintain the hope for a new heaven and a new earth. The realisation of this hope is already promisingly announced in God's Trace in nature itself. Catherine Chalier has accordingly led Emmanuel Levinas' alterity thinking along ecological lines. The Scripture says man is the last step in creation. The entirety of God's creation was already there before man was created. Man, in other words, has to 'discover' the world, and can never pretend to be the source and origin of everything. We are discoverers ('exegetes') before we are creators ('genetisists'). Before we are to rule, we find ourselves in a relationship of givenness. 'Having dominion' is not the first step, man is placed in a certain relation to nature. Man is not only the *last* creation but also the *first* to be punished.[11] This demonstrates how man's relation to the creation is to be understood: as an ethical relationship.

Finally, let us briefly return to Levinas' reticence towards nature. Although nature is God's good creation, it eventually also has a threatening meaning. An aesthetic, holistic harmony model of nature is naïve, because it insufficiently takes into account nature's threatening disposition for man. The aids virus does not deserve any

[11] Roger Burggraeve, "Scheppingsvisie en ecologische opdracht: perspectieven vanuit Genesis 1," in Axel Liégeois, Joseph Selling, Luc Anckaert, Johan De Tavernier, Bert Roebben, Johan Verstraeten, *Aspecten van een christelijke sociale ethiek* (Louvain: Library of the Faculty of Theology, 1991), pp. 125-136.

kind of respect. Genesis says that man 'has dominion', and this is also exactly what *should* be said (contrary to 'ecocentrism'). This phrase does not only oppose a certain (subjugating) God concept and does not only exalts man to a position of importance, but also reflects the experience of nature's ambivalence. As such, it is not so much the question whether man's return to nature is important, but rather the question to *which* earth we should return. 'Earth' with a capital is too good to be true. With Chalier, we have opted for an anthropologically understood biblical-ecological revival – albeit not for any kind of anthropocentrism, but rather for an *ethically qualified* anthropocentrism. 'Which also means that God's great work waits for its exaltation by man.'[12]

[12] Concluding sentence in Chalier, O.c., p. 207.

In Response to Didier Pollefeyt

David Patterson

Since I am generally in agreement with Didier Pollefeyt's thesis, my intention here is to shore up his argument and make it even stronger, as well as to offer a critique. One way to do both is to provide a slight correction to some of the claims that Pollefeyt makes concerning Jewish tradition in the Diaspora. One assertion that he makes, for example, is that since the Jews' expulsion from the Holy Land at the hands of the Romans, "the great Jewish liturgical festivals, which commemorate historical events, were celebrated without reference to the moments in nature's cycle with which they coincide and which they factually celebrate." This is not exactly the case.

The observance of *Sukkot* is a case in point. This holy season is celebrated by dwelling in "booths," with a makeshift roof made of natural greenery, for seven days in the fall; it is an observance consciously linked with a sense of being in the midst of nature, as God's guests in His natural world,. In the *musaf*, or 'additional prayers,' for *Shemini Atzeret* (the day after the sevens days of *Sukkot*) Jews recite the Prayer for Rain; in the musaf for the first day of Passover (in the spring) they recite the Prayer for Dew. And in their daily prayers they include the affirmation of God as the source of rain in the winter and the source of dew in the summer in their daily prayers. In addition, there is the liturgy of the Sanctification of the Moon, which is typically said at the conclusion of Yom Kippur, but it can be said in any month, starting seven days after the first of the month, which is itself measured according to the lunar cycle. Therefore even in the Diaspora the Jews have maintained the sensitivity toward the cycles of nature that Pollefeyt correctly believes to be essential to a Jewish relation to the natural world.

Pollefeyt further asserts, quite correctly, that in the Jewish tradition "man is called to uncover and unravel the Traces that God has left in his Creation, and to bring new life to their meaning. Yet, this presupposes a hermeneutical attitude towards nature on the part of the believer." In fact, he is even more correct than he may realize. The *kabbalistic* tradition teaches that all of creation, including nature, is imbued with divine sparks, which consist of the divine Word that continually brings all things into being. Nature is saturated with the Word, and, broadly speaking, God's Word is Torah: nature is full of Torah and teaching and divine utterance, which we must decipher. In Judaism, then, there is a long-standing tradition of what Pollefeyt calls "a hermeneutics of nature." The contemplation of "that other great riddle: creation" is central to the Jewish mystical tradition, which plays a very significant role in Judaism in general and in Chasidism in particular, as Pollefeyt correctly points out. Indeed, when the Midrash declares that Torah is the blueprint for creation (Bereshit Rabbah 1:2), it implies the very thing that Pollefeyt ascribes to Catherine Chalier: "In Chalier's thought, the Scripture thus is the necessary mediation between man and nature." Strictly speaking, the Word by which nature has its existence is Scripture.

One of Pollefeyt's most profound insights comes with his reading of the Book of Job, where he notes, "God just presents Job with His Creation as if His answer to Job's misery lies there." But if he had the space, Pollefeyt could pursue the ramifications of his insight a bit further. It might suggest, for example that the problem of evil that Job raises – namely, the question of why the righteous suffer and the wicked prosper – lies in a view that would reduce nature to the stark indifference that Levinas associates with the 'there is.' If Torah speaks from the depths of nature, then Torah cries out, with Job, over injustice in the world. Why? Because, just as injustice does harm to the soul, which is made of Torah, so does it do harm to nature itself – and to nature's design, which rests upon the Word that continually sustains its existence. The transformation of nature that will come in the messianic age will come as a result of the mending of Torah: if the wolf does not yet lie down with the lamb, nor the leopard with the kid, it is because we have not yet attained the spiritual mending that will transform the natural world. Because everything, both within and without, is connected by Torah, everything spiritual affects all things natural, and vice versa. This is one implication that Pollefeyt might pursue further.

Another implication that he might pursue further lies in his comments on the desert versus the land of milk and honey. First, a bit of a corrective from the Jewish tradition is called for: the Midrash says the Israelites were reluctant to enter the land precisely because God had provided for all their needs in the desert (see, for example, Bamidbar Rabbah 16:24), whereas in the Land they would have to till the earth and work for their sustenance: they would have to build something, a dwelling place for the Holy One in the midst of His creation, as symbolized by the Temple. The lesson is that this engagement with nature is an essential part of entering into a deeper relation with the creator of nature; it puts us in a position of snatching our hard-earned bread, rather than the unearned manna, from our own mouths and offering it to another. This offering to another on the part of a dis-interested self comes as a result of our dominion over nature: here lies the opening that enables God to become manifest in the natural realm that he already permeates.

Other ramifications that Pollefeyt might pursue in the future derive from his critique of a modernity that has robbed nature of a voice that speaks to the human heart and soul. He might consider, for instance, how modernity has emerged from a fundamentally Greek ontological outlook. Influenced much more by the Greeks than by the Hebrews, modern science, at best, can do no more than posit a First Cause, an It, such as a Big Bang, and not a Creator, a Who, with whom we live in a fundamental, definitive relation as those who are created in His image—and not evolved from Its ooze. Only where God is understood as Creator can we determine an anarchic tie to one another, in Levinas's sense of the anarchic, where each soul, as an emanation of the Creator, is tied to the other through its tie to the Creator. Only if we can determine this tie to one another can we determine an anarchic ethical responsibility either to nature or to each other. Pollefeyt grasps this point, inasmuch as he suggests that without the ethical relation to nature, there can be no ethical relation to the other. But one question he leaves unanswered in this connection is this: Has Levinas, then, failed? Given his dismissal of an ethical relation to nature, is Levinas's thinking about ethics and infinity inadequate to the point he wants to make about ethics as a first principle? If so, has he failed to answer Heidegger, the Nazi champion of ontological, totalitarian thinking, who was also a voice for the pagan worship of nature that Pollefeyt rightly condemns?

Which brings me to one last concern. Pollefeyt has made excellent use of teachings from the Jewish tradition to make his point regarding a post-Holocaust understanding of our ethical obligation to the natural world. After all, the Nazis set out to destroy the very teachings and traditions from which Pollefeyt draws so heavily. However, just as he overlooks some of the ontological tradition's contribution to the desecration of nature, so does he leave out anything that Christian tradition might contribute to his insistence on our ethical obligation to nature. But does the Christian tradition have anything to contribute? For a renowned Catholic theologian such as Didier Pollefeyt, the elephant in the room is this: to what extent has the Christian *contemptus mundi*, which is alien to Jewish teaching, contributed to a desecration of the natural, material world? And, just as crucial to his argument, to what extent might the Christian contempt for the material world have contributed to a Christian indifference toward the gassing and

burning of the body of Israel under the Third Reich? It seems to me that these two phenomena are related.
I am not sure it would be enough to invoke something like St. Francis of Assisi's 'brother sun, sister moon,' given the scriptural call for such things as celibacy (see 1 Corinthians 7:7-8). This view of the body – a divine creation of nature – as the enemy may well be the result of a corrupting Greek influence (see, for example, Plato's Phaedo). If so, that influence, which also influenced modern science, should be addressed. In any case, I think there is much more to be explored in this connection. It is a needful elaboration on Pollefeyt's excellent insights.

In Response to Didier Pollefeyt

Sarah Pinnock

Is the Bible an obstacle or a guide in the ecological debate? Roman Catholic theologian Didier Pollefeyt takes a philosophical approach to the Bible guided by the framework of Jewish thinker Catherine Chalier, a former student of Emmanuel Levinas. While his title highlights the role of the Bible, the chapter itself does not delve into specialized exegesis or historical-critical textual studies. Rather, he develops a hermeneutics of the Bible and nature with wide ranging scope, an approach that raises multivalent questions. Intrigued by his proposal, my response will unpack key aspects of his argument and its underlying assumptions.

Genesis 1 is an important text for Pollefeyt. He observes that the seven day creation narrative depicts nature's cycles as a miracle, where God creates reality at every moment. Even in Nazi camps, he insists that the divine 'miracle' of nature was perceived. As he writes, "Many Jews in Auschwitz also continued their prayers and the celebration of the liturgical holidays, wherein, even more than before, the connection with nature's cycles played an important role." (148) These examples authorize post-Holocaust thinkers to affirm God's traces in nature, rather than give Hitler a posthumous victory by viewing nature as faceless and cannibalistic, i.e. by denying God's reality in creation.

However, Pollefeyt also considers how Genesis 1 raises problematic ecological issues. When God creates humans in his image, he gives them dominion over all of the fish, birds, animals, trees, and plants. God blesses humanity and instructs them to populate the earth and subdue it (Gen 1:26-30). I would add that even more prominently in the Garden of Eden narrative (Gen 2:4-3:24), domination and subordination permeates creation where nature and woman are subordinate to man, and God has dominion over both human beings and nature. Feminist scholars, among others, note that there are inherent problems with 'power over' as a hierarchical model for the divine-human-nature relationship. A suggested alternative is a cooperative model of power as 'power with' in mutuality between God and humanity, in other words, a non-hierarchical model of power.[1] Although he recognizes the damaging social consequences of human dominion

[1] Dorothee Sölle, *Thinking about God* (Philadelphia, PA: Trinity Press International, 1990), p. 51.

over nature, his essay concludes by maintaining a hierarchical (ethically qualified) anthropocentrism which rehabilitates the notion of 'dominion' drawn from Levinas' ethics of alterity. More could be done to confront the full extent of the obstacles arising from the creation narratives.

On a philosophical level, Pollefeyt launches a critique of Emmanuel Levinas, whose writings are tremendously influential in post-Holocaust discussion. To paraphrase his title, his chapter poses the question: "Levinas in the ecological debate: obstacle or guide?" With respect for Levinas, Pollefeyt finds his ethic lacking ecological sensitivity. The major problem lies in how Levinas approaches nature as '*il y a*' (there is), an impersonal environment surrounding human individuals. Humans struggle against *il y a*, which is faceless, alien, and indifferent to persons. Biographically, Pollefeyt suggests that Levinas' internment in a *Wehrmacht* camp in 1939 and the internalized trauma of the Holocaust, may explain this estrangement from nature. On the other hand, Holocaust experience also accounts for a strong point in his ethics, namely, emphasis on the vulnerable face of the other, and the commanding responsibility that the other places on the self. Since it is usually presumed that Holocaust reflection fosters comprehensive moral awareness, it is rather startling to consider that attention to the Holocaust may actually eclipse attention to nature.

However, perhaps it is not the Holocaust that explains Levinas' lack of an ecological ethic, but rather the tradition of Jewish ethical reflection. Pollefeyt observes that far from being isolated, Levinas' standpoint actually reflects a pattern of Jewish forgetfulness toward nature, explained by historical factors such as rejection of pagan idolatry, exile from Israel, and diasporic life among foreign peoples. Nevertheless, there are exceptions such as Hasidic mysticism that honors nature as a manifestation of God, as well as Neo-Platonic Jewish philosophies of Philo, the *Zohar*, and the *Lurianic Kabbalah*, as well as attention to animals and land in the Hebrew Bible. Affirming nature, the seasons of fall and spring, which end and begin the growing season, evidently inform the symbolism of holy days such as *Rosh Hashanah*, *Sukkoth*, and Passover.

As a corrective to Levinas, Pollefeyt praises Catherine Chalier who extrapolates Levinas' thought in an ecological direction. She speaks of the 'Traces' of God found in nature, as a counterpart to Levinas' claim that God reveals himself in the face of the other. This claim navigates between two extremes: nature is neither divinized, nor desacralized. Moreover, nature is not *il y a*, mute and alien. The discovery of God's traces in nature, like scripture, is a hermeneutical process. I agree with Pollefeyt that nature reveals God. But he is careful to insist that scripture is "the necessary mediation between man and nature." (144) He warns that contact with nature reveals its immanent beauty only, insufficient in itself and potentially equivalent to pagan idolatry. He maintains biblical authority over disclosure of God's traces in nature.

Employing the Bible as a positive guide, Pollefeyt finds the book of Job indispensable. To explore how nature reveals the divine, he focuses on God's climactic speeches to Job from the whirlwind. Pollefeyt observes that Job's suffering generates his distress and confusion. More specifically, Job's suffering blocks his awareness of nature. As he puts it: "The hermeneutical paralysis of suffering men is a consequence of their inability to safeguard some space in themselves where the other can be received. Job's immense sufferings deprived him of the ability to look to nature through a different frame of mind." (145) Job's suffering and Levinas' Holocaust experience both accompany alienation from nature. Yet it is precisely a new perception of nature that turns Job around.

God answers Job out of the whirlwind with speeches that show the beauty of nature to someone in misery. As God shows the limits of human knowledge, Job's focus on justice is supplanted by attention to natural majesty and divine power. Job discovers traces of God in snow and hail, goat and ox, and dangerous creatures. Rather extraordinarily, given Job's previous questions and complaints, God's manifestation to Job is nature-centric rather than anthropocentric, and aesthetic rather than ethical. But particularly in view of the Holocaust, I am concerned that Pollefeyt ignores the radical implications of this epiphany.[2] Before the ferocity of storms and beasts, human beings are weak and pitiful. Behemoth and Leviathan symbolize natural forces of evil, over which God rules. With such divine power, no purpose of God can be thwarted. But awesome nature with its fierce creatures is threatening even unto death, like the *il y a* identified by Levinas. The harsh beauty of nature under God's command raises serious theodicy questions.

Based on the whirlwind speeches, God's intentions are not comprehensible in the affairs of human life measured by human justice. God intimidates Job and subdues his questions, which results in pious but groveling submission. After the Holocaust, questions about God's rule over creation and fairness to the righteous are exacerbated by the natural majesty manifest in God's speeches. The pagan

---

[2] I consider the insufficiency of the book of Job as a theodicy through reflection on the interpretations of Martin Buber and Ernst Bloch in Sarah K. Pinnock, *Beyond Theodicy: Jewish and Christian Continental Thinkers Respond to the Holocaust* (Albany: SUNY, 2002), pp. 71-74.

'pantheism' that Pollefeyt repudiates, associated with Martin Heidegger and Richard Rubenstein, involves a God that exercises "fascinating and frightening attraction" and a numinous power "that dissolves the basis for any kind of personal response-ability" (142). But arguably, this description resonates with Job's encounter with the God of the whirlwind. Is the God of Leviathan a God who confers moral responsibility upon humans to enact justice? In view of these considerations, the book of Job presents an obstacle, and not only a guide to ecological hermeneutics.

Connected to Job's vision of nature, Pollefeyt highlights the Dutch Jewish diarist Etty Hillesum as a Holocaust example of the appreciation of nature, despite suffering. He points out that even in the Nazi camp of Westerbork, Hillesum is inspired by the beauty of blooming jasmine flowers. This example of a 'desert experience' (Chalier's term) is meant to parallel Job's discovery of God's traces in creation, although the fearsome quality of the whirlwind is absent in Hillesum's situation. But does scripture play the role of necessary mediation toward discovering God in nature? From her diaries, we know that as a nonobservant Jew, Hillesum's perception of God arises from many different textual sources and personal influences. She finds God mediated in literature and poetry, for instance, in the works of Tolstoy, Dostoyevsky, Augustine, and Rilke whom she adored.[3] I observe that her perception of God in nature is not primarily a hermeneutics based on scripture. In comparison to Hillesum, an assimilated Jew, Job's relationship to the people of Israel is also somewhat ambiguous for the Bible tells us only that he is a righteous man in God's eyes. Rather than through understandings of Torah, Job meets God through direct relation and encounter. I do not dispute that Hillesum and Job discover God's traces in nature, or that their discoveries may be consistent with scripture. However, their examples do not confirm Pollefeyt's assumption about the necessary authority of scripture for a hermeneutics of nature. If nature and scripture are different languages to be interpreted, God's traces may be revealed not only by scripture but also by other religious writings, poetry and literature, and direct encounter with God.

The biblical texts that Pollefeyt features near the end of his chapter depict the prophetic vision of God's Kingdom. The Bible is a guide in imagining the restoration of humanity and nature, particularly the prophets Ezekiel and Isaiah. In this utopian vision, the effects of sin will be reversed, reconciliation will occur

---

[3] Etty Hillesum, *An Interrupted Life: The Diaries of Etty Hillesum 1941-43*, Jan Geurt Gaarlandt trans. (New York, NJ: Washington Square Press, 1985), p. 220.

among all people, and between humanity and nature. 'Poisonous snakes will not harm children, nor will lions hunt lambs' (Is. 11:6-9). Pollefeyt refers to messianic peace. However, he does not tackle differences among various Jewish and Christian interpretations of the Messiah. Nor as a Catholic theologian does he address the spiritualization of God's Kingdom achieved by salvation through Jesus Christ, found in Pauline epistles and other church writings. Such non-nature-based interpretations of the messianic age drastically demote the importance of the material world.

At the end of the chapter, Pollefeyt employs philosophical reflection on the distinction between human beings and animals, using the anthropology of alterity found in Levinas and Chalier. In so doing, he returns to the creation narrative in Genesis 1:26 as a guide. He remarks that if humans have self distancing ability and concern for the larger good, then dominion over nature needs not to destroy or exploit. But in contrast with earlier positive emphasis in the chapter on natural beauty and God's traces, Pollefeyt concludes with the negative observation that nature often threatens human life and must be held in check. If we do not idealize natural harmony naively, we will recognize that nature's ambivalence warrants human guidance. In the end, it seems that Levinas might be correct in asserting that nature is threatening, yet paradoxically, nature also manifests God as Chalier insists.

Impressed by Pollefeyt's complex proposal, I shall end by identifying unresolved issues for Jewish-Christian dialogue. Pollefeyt clearly accepts Jewish understandings but he does not relate them explicitly to Roman Catholic doctrine. His approach centers on the Hebrew Bible, on Jewish philosophers, and on Jewish Holocaust experiences. What about the role of the New Testament as obstacle or guide to nature? How might Pollefeyt engage Christian doctrines of creation and God's Kingdom in relation to Jewish thought? It would be interesting to examine notions of God's traces among other Jewish and Christian writers as well as in Chalier's writings. Lastly, I am left wondering how the threat of nature displayed by Levinas and the amoral beauty of nature displayed in Job are reconciled with divine justice and ethics after Auschwitz.

In Reply to My Respondents

Didier Pollefeyt

David Patterson at the same time affirms, corrects, radicalizes and challenges my analysis of the Jewish understanding of the human being as a guest in the natural world created by God. David Patterson explains how for the Jewish tradition nature is saturated with divine sparks coming from the Word of God which continuously recreates natural and human reality. The idea of 'Nature as full of Torah' makes my analysis much richer. By reflecting on the book of Job, he explains how the idea of the 'Torah as a blueprint for creation' implies that injustice to nature, as injustice to man, is a violating of the Torah and thus in contradiction to the will of God. And put positively, that a mending of the human world cannot be disconnected from a mending of nature. This is illustrated by the idea that the Jewish people could only enter into the Promised Land, and so into a deeper relationship with God by an engagement *vis-à-vis* the land and by sharing its limited resources with the other. I am also in agreement with David Patterson that modernity with its ontological categories has silenced and suffocated the Word of God in nature and has disconnected human beings from each other and from the earth. Levinas is therefore so crucial, because he has put ethics again as the first principle in front of contemporary philosophy. Has Levinas failed with his dismissal of an ethical relation to nature, David Patterson asks, especially in light of Levinas' ongoing discussion with Heidegger and his pagan worship of nature? I think that Levinas has, together with the Jewish tradition, presented a sound word of caution *vis-à-vis* a nature whose powers can easily be divinized. Nazism is a clear illustration of the fact that respect for nature can go together perfectly with disrespect for human beings and how a pagan definition of nature played a constitutive role in its deadly ideology. What I want to underline in my chapter is that Levinas' Jewish philosophy has a *potentiality* to value nature without ending up in a pagan divinization of nature as such, but that this potentiality remained undiscovered in Levinas' own thinking. I think this is understandable in the light of both his personal and existential confrontation with the brutal forces of nature during his imprisonment under Nazism and of his philosophical wrestling with Heidegger's philosophy, and more broadly with ontological and totalitarian thinking as such. With Catherine Chalier I believe that Levinas' thought has not failed at this point and that it can be made relevant and meaningful for contemporary ecological concerns. Chalier further developed Levinas' philosophical framework based on an anthropocentrism that is ethically and theologically qualified, meaning that it is oriented by an ethical monotheism: the belief in a personal and commanding God who gives power to man to rule over creation in a moral way. It is precisely this monotheistic framework that is challenged by my other respondent, Sara Pinnock. As is also clear in other places in this volume, Pinnock argues rather for a direct experience of God in nature, or identifies God with the blind forces of nature itself. She criticizes from a feminist perspective the idea of humanity having 'power over' nature because this model still reflects a hierarchical model inspired by male dominance. Instead she opts for a more cooperative, non-hierarchical model of 'power with', a mutuality between God and humanity. In this context, Pinnock resists the idea of God who reveals himself (only) indirectly in the traces he left in his creation. She suggests that the book of Job is not only to be read as support for a hermeneutical understanding of our relation to nature, but that it also forms an obstacle to this view. God shows his divine power directly to Job in and through nature. God's answer to Job is "nature-centric rather than anthropocentric", "aesthetic rather than ethical". In Pinnock's interpretation, God appears to Job not as an ethically consistent God but as a natural force that intimidates Job, forcing him to total submission and silencing his moral protest against the injustice of his suffering. Etty Hillesum found God in nature but Pinnock minimizes the mediating role of scripture in Hillesum's religious experience referring to the fact that she was a nonobservant Jew and to the role of profane literature in her writings. In this way, Pinnock questions Scripture as a necessary medium for encountering God in nature. "God may be revealed by other texts and by a direct encounter with God", Pinnock concludes. My problem with Pinnock's approach is that she turns "humanity's power over nature" into "nature's power over humanity". She turns the male-female power structure upside down but does not overcome it, on the contrary. Levinas made us aware of the depersonalizing, 'il y a-tic' power of nature, a power with no compassion, a power with no moral sensitivity. That is the reason why in my view humanity *should* dominate over nature. Nature is not just a romantic place. This control over nature is not just a male enterprise, but a human vocation, serving both men and women and especially their children. This domination should not be inspired per se by a 'will to power', but should be understood in the line of Levinas as an ethical responsibility for something that is given to us as a gift and that still bears the traces of its divine giver/creator. Even if nature in this view is not a moral subject itself, neither human nor divine, it is an object of our moral consideration because it is a divine creation. One of the lessons to be learned from Nazism is how dangerous a divinization of natural powers can be. It was Nazism that submitted itself and the whole world to the divine powers of nature as they understood them. In this sense, I fully agree with Fackenheim when he criticizes Rubenstein and his mystic nihilism ('God as Mother Nature') as a posthumous victory for Hitler. It is true that the eclipse of nature in Levinas' thinking goes back to a much longer tradition of Jewish ethical reflection, as Pinnock rightly points out, but the holocaust makes Levinas' attitude towards nature and the divinization of the blind forces of nature even more understandable and necessary. In the book of Job, it is true, God speaks through the manifestations of nature, but Job resists. But the book of Job is protest literature. The message of the book of Job is precisely that Job becomes heated and angry *vis-à-vis* such a hostile and morally incomprehensible deity. One cannot isolate some quotes of God or attributed to God from the book of Job and its central message. Especially in the epilogue of the book of Job (written by another author), God himself says that Job is in the right (Job 42:7). I do agree with Sarah Pinnock that the book of Job poses the question of theodicy, that is the relation between God and human suffering, but theology has provided many other answers to this question than by putting evil into God himself who delivers humanity over to pure arbitrariness (*cfr. infra*). It seems to me that the Jewish-Christian tradition has liberated itself progressively from such a kind of immoral God. Both Patterson and Pinnock challenge me as a Christian, more specifically as a Catholic theologian on these issues. I am in agreement with David Patterson (and Sara Pinnock) that, under the influence of Greek thinking in the course of the centuries, Christianity developed a very negative view of the human body, and more generally, the material world. I do believe that this is an element that historically contributed to the Christian relativisation and legitimization of the drama and the evil of human suffering, especially in relation to the gassing and the burning of the body of Israel during Nazism. From this perspective, the topic of holocaust and nature also presents a challenge to Christian/Catholic theology and to the Christian/Catholic tradition as such. A post-holocaust understanding of nature should avoid a devaluation of the material world, of the body, of creation. One could indeed refer to St. Francis, as David Patterson suggests, but the answer for Christians is to be found more in the center for their faith, namely in the events of the incarnation and resurrection of Jesus Christ. Incarnation means that the Word became Flesh, that God became material in the world. Even more, God made himself vulnerable to the world and suffered through his Son on the cross. This is another way of looking at the question of theodicy: in a story where God suffers, suffering can no longer be turned against God; it is a risk connected with freedom and a consequence of evil. Resurrection, especially the resurrection of the body, means for Christians that the whole human person and not just an abstract soul will be saved. Especially in the Catholic tradition, and in Catholic liturgy in particular, sensitivity to the body, the material, to a connection with the seasons of nature plays a central role. For a Catholic, it is not difficult to see and to value that people – like Etty Hillesum – can experience God outside scripture as a medium. Catholicism is not based on a protestant idea of *sola scriptura* ('scripture alone'). Even if Etty Hillesum was very familiar with the bible and with the non-religious authors who inspired her too, the Catholic tradition can recognize that other traditions and human experiences can also reveal aspects of God, since there the Word of God (the Logos) is also at work (even if this work of the Logos can never be disconnected from Christ). This further explains why God can be experienced without the mediation of scripture (even if the fullest meaning of life for Christians can only be found in the light of the gospels). For Christians the incarnation and resurrection imply that God engages with humanity and the world, fully and in a unique way, and that God will save the whole person, body and soul, individual and community, culture and nature. This stands in radical contrast with the pure spiritualization of the Kingdom of God that was part of the working history of Christianity and that often led to a dualistic and anti-natural understanding of salvation. Christians live in the hope of the liberation of the whole creation. In contrast to popular presentations, Christianity does not claim that we will forever stay in heaven, but rather in a newly transformed material world, a new heaven and a new earth (Revelation 21:1). Saint Paul speaks clearly in Romans 8 about the destiny of the natural world as not being one of destruction, but of transformation: "the creation itself will be liberated from its bondage to decay and brought into the freedom and glory of the children of God" (Romans 8:21).

Vulnerable Bodies

Feminist Reflection on the Holocaust and Nature

Sarah K. Pinnock

Nature is commonly defined in counterpoint to the human. In contemporary society, we utilize nature for two main purposes: as material support for demographic and economic growth, and for aesthetic and recreational enjoyment. Belatedly, there emerged recognition of multifaceted crises facing nature, due to human exploitation, when Western governments began to implement environmental protection programs in the 1970s.[1] But it was earlier in the twentieth century that the stark crisis facing nature already emerged clearly. In World War II, advances in science created unprecedented conditions of mass death employing tanks and artillery, aerial bombardment, nuclear explosions. Moreover, the first industrial genocide used the professional expertise of doctors, scientists, and administrators, to engineer deportations, gas chambers, and concentration camp labor. On a shocking scale, Holocaust history displays the manipulation of nature to create technologies of death and the devaluation of nature.

Ethically, the Holocaust is considered a paradigm example of evil. As discussed by classical Christian authors such as Augustine, evil is divided into two types: natural and moral. Natural evil is caused by powerful forces like earthquakes, hurricanes, forest fires, floods, diseases, and dangerous animals, while moral evil is caused by humans. Popular fascination with Adolf Hitler and Nazi perpetrators, as well as ordinary people, who followed Nazi orders, indicates how the Holocaust provides fascinating cases for reflection on moral evil. But what about nature? Did natural evil play a role in the Holocaust? Diseases, storms, and winter cold, among other natural evils, claimed lives on all sides of the war. Nevertheless, natural causes of death were dwarfed by human killing, and nature was itself a victim. But in many ways, this conceptual dualism between moral and natural is itself problematic. It masks the decisive human contribution to natural evil, connected with the technological capacity to prevent harm and alleviate suffering. With modern communication, transportation, and medicine, there is little clear cut distinction between moral and natural evil. Certainly, it is trou-

[1] For instance, the US federal Environmental Protection Agency was founded only in 1970.

bling that the divide between what is moral and natural potentially makes nature seem irrelevant to Holocaust reflection.

This essay approaches the Holocaust and nature guided by feminist interests and informed by women's perspectives. At the outset, some self-disclosure is appropriate to situate my thinking. Academically, my specialization is philosophy of religion with particular interest in the problem of God and evil.[2] As a dual citizen, I grew up in Canada and moved to the United States to pursue a doctorate at Yale and a career as a religious studies professor. My religious background is Protestant, mainly Episcopalian, and my interpretation of the Christian faith is feminist and liberation oriented with emphasis on the prophetic and mystical dimensions of faith.[3] Experientially, I have found my thinking enriched by interreligious opportunities including Buddhist meditation and Jewish-Christian-Muslim trialogue. I am personally and professionally committed to reflection on the Holocaust as a historical, religious, and moral crisis.

WOMEN, NATURE, AND THE HOLOCAUST

There is a long standing association between women and nature in Western thought. From classical Greek philosophy onwards, a basic dichotomy is posed between men and women parallel to dualisms between form and matter, mind and nature, reason and emotion, activity and passivity. For instance, according to Aristotle, the male is the active agent in procreation who supplies the form and the impulse for development, while the female is passive in supplying the matter for the unborn child and incubation conditions.[4] Aristotle claims that male have unique capacities of reason, superior to women, based on the analogy of the soul's superiority over the body. Until the nineteenth century, at least, this hierarchy between men and women was scientifically justified by biological reasoning. This prejudicial interpretation of sexual difference has pervaded philosophy, theology, and science, as well as popular attitudes. Of course, critics of this para-

---

[2] My published dissertation on Holocaust responses deals primarily with four authors: Gabriel Marcel, Martin Buber, Ernst Bloch, and Johann Baptist Metz. Sarah K. Pinnock, *Beyond Theodicy: Jewish and Christian Continental Thinkers Respond to the Holocaust* (Albany, NJ: SUNY, 2002).

[3] For one exemplar of mystical prophetic theology that I admire, see the essays collected in: Sarah K. Pinnock, ed. *The Theology of Dorothee Soelle* (Harrisburg, PA: Trinity Press International, 2003).

[4] Aristotle, *Generation of Animals,* trans. A.L. Peck (Cambridge, MA: Harvard University Press, 1953), 765b, pp. 9-16.

digm, including liberal feminists, object to associating women with nature and the body, and emphasize men's and women's common faculties. Still, by categorizing men and women together as human, and nature in opposition, leaves the basic subordination of nature unchanged.

Since patriarchal assumptions continue to operate, the association between women and nature is reinforced. Women's bodies receive attention as sexual objects to be appreciated, enjoyed, and possessed by men. By virtue of their bodies, women are objectified and subordinated. Moreover, women traditionally have domestic responsibilities associated with material tasks such as cooking and cleaning, buying food and clothing, and tending children that continue today. This practical role in the household arguably puts women in closer contact with natural processes, which may enable distinctive awareness of nature and ecological issues, such as the ill effects of pollution, food additives, or chemical waste. As caregivers for children and the elderly, often, a woman's perspective places social problems in a distinctive light.

Rather than being a liability, viewed positively, the association between women and nature highlights women's uniqueness. Cultural feminists embrace the material and emotional and bodily side of Aristotle's dualisms as a platform from which to criticize patriarchal assumptions and develop appreciation of women's caring roles. Biologically, women's bodies are equipped for pregnancy and lactation. If it is accepted that thinking is never entirely detached from embodiment, then perhaps women's domestic roles and reproductive potential might make the connection with nature fruitful, rather than oppressive. Nevertheless, categorizing women alongside nature may merely mimic patriarchal constructs and risks perpetuating conflicting polarities. The fact remains that the estrangement of nature from 'man' is a harmful pattern of thought that has proven morally deficient, both towards women and what is non-human.

The Holocaust manifests chronic symptoms of this conceptual devaluation of nature and the body in Western thought. The harmful repercussions of modern science and technology displayed by the Holocaust and World War II have been explored by authors such as Martin Heidegger and Jaques Ellul. More recently, the Jewish philosopher, Edith Wyschogrod, has traced how mass death resonates with major philosophical themes from classical Greek to modern German philosophy.[5] In this essay, I shall consider critiques and alternative ways of thinking

---

[5] Connections between the history of Western thought, technological abuse of nature, and the phenomena of mass death have been traced by a number of philosophers, including

about the Holocaust and nature using women's perspectives. My interests center on feminist concerns about subordination and hierarchy, and ideals of reciprocity and mutuality. I warrant that a holistic approach to humanity, nature and God can provide critical purchase on faulty dualisms, ethical insight, and constructive possibilities for religious reflection.

Gendered approaches to the Holocaust are a relatively recent phenomenon.[6] Since research on women in the Holocaust accelerated in the 1990s, there have been hesitations about such scholarship, including sharp criticism by prominent Holocaust scholar Lawrence Langer.[7] His most persuasive concern, in my mind, is that woman-focused approaches might detract from the important fact that "Jews" were all victims, male and female. He betrays anxiety that feminist scholars might declare women the chief victims under patriarchal oppressors, or create a moral hierarchy by praising female virtues such as caring and nurture as crucial to survival. To repudiate such gynocentric flattery, Laurence Langer draws on women's audiovisual testimonies from the Fortunoff archive to show that women manifested non-ideal traits. Women in the Holocaust sometimes betrayed loved ones, became isolated, acted selfishly, and lost hope. There is no uniform manifestation of caring that applies to women in general, and thus, no morally unique reaction to Holocaust suffering.[8] Although I agree with his conclusion, I find Langer unnecessarily defensive in his insistence that women exhibit character flaws under genocidal pressure. He misinterprets feminist interest in what women have to offer, experientially and intellectually, as a glorification

---

Martin Heidegger, *The Question concerning Technology*, trans. William Lovitt (New York, NJ: Harper and Row, 1977); Jaques Ellul, *The Technological Society*, trans. John Wilkenson (New York, NJ: Random House, 1964); and Edith Wyschogrod, *Spirit in Ashes: Hegel, Heidegger, and Man-Made Mass Death* (New Haven, CT: Yale University Press, 1985).

[6] For examples of scholarship on women and the Holocaust, see John K. Roth and Carol Rittner, *Different Voices: Women and the Holocaust* (New York, NJ: Paragon House, 1993); Judith Tydor Baumel, *Double Jeopardy: Gender and the Holocaust* (London: Vallentine Mitchell, 1998); Brana Gurewitsch, ed., *Mothers, Sisters, Resisters: Oral Histories of Women Who Survived the Holocaust* (Tuscaloosa, AL: University of Alabama Press, 1998).

[7] John K. Roth, "Equality, Neutrality, Particularity," in Elizabeth R. Baer and Myrna Goldenberg, eds., *Experience and Expression: Women, the Nazis, and the Holocaust* (Detroit, MI: Wayne State, 2003), p. 10.

[8] Lawrence Langer, "Gendered Suffering? Women in Holocaust Testimonies," in Dalia Ofer and Lenore J. Weitzman, eds. *Women in the Holocaust* (New Haven, CT: Yale, 1998), p. 361.

of women and corresponding denigration of men. In this essay, my focus on female authors is motivated by desire to include more women's voices in dialogue, and not to create gender dichotomies. I am convinced that women's experiences and feminist perspectives are broadly relevant to Holocaust reflection.

WOMEN'S INTELLECTUAL RESPONSES

This essay deals with the work of two Jewish thinkers whose writings relate directly to nature and the Holocaust: French philosopher and mystic Simone Weil (1909-1943), and Jewish feminist Melissa Raphael, author of *The Female Face of God in Auschwitz: A Jewish Feminist Theology of the Holocaust* (2003). They write at different times, with different interests, and by no means converge on one singular women's perspective. In fact, it intrigues me that I find richness in both authors given their separation. For Raphael as well as Weil, who predates feminism, central issues in feminist theory emerge: embodiment, reciprocity with nature, divine immanence, and resistance to oppression. This essay is an experiment, offering a preliminary investigation of the constructive possibilities found in the work of these two authors. It centers on the contributions made by Weil and Raphael toward exposing problematic attitudes towards nature and developing more responsible perspectives.

SIMONE WEIL

As a French Jew, a Marxist socialist, a philosopher, and a mystic attracted to Roman Catholicism, Weil is unique. Her education in Paris at prestigious institutions taught her appreciation of Hellenic, Christian and French culture, and generated passion for modern philosophy. Weil is Jewish, but her family was nonreligious and she considered her Jewish identity as a genealogical fact.[9] She is famous for embracing Roman Catholic piety, but refusing baptism with carefully reasoned arguments. She also found religious inspiration in non-Christian sources such as the Bhagavad Gita and classical Greek writings. Weil fled with her family to the south of France in 1940, immediately before the Nazis invasion

---

[9] Like many assimilated French Jews, Weil viewed Judaism pejoratively. Her negative view of Judaism blocked identification with Jews during the Holocaust, rather, she identified with victims of the war in general. Thomas Nevin, *Simone Weil: Portrait of a Self-Exiled Jew* (Chapel Hill, NC: University of North Carolina Press, 1991), p. 251.

of Paris. In 1942, concerned about her parents' safety, Weil journeyed with her family to the United States, where she remained for only four months. On moral compulsion, she traveled alone from the US to England where she could work more directly with the French resistance, but her health deteriorated, and she died suddenly in 1943. Even in her last months, she was actively looking toward the future, and feverishly engaged in writing *The Need for Roots*, a book analyzing the distortions of Nazi ideology and the possibilities for rebuilding European society after the war. In the following paragraphs, I shall briefly trace her views on nature, force, embodiment, and relations between self, nature, and God.

Weil condemns the immorality of modern science and technology. It is clear to her that technology, instrumentalization of nature, and warfare are closely linked. In *Oppression and Liberty*, a collection of essays written during the 1930s, she vividly describes the distorted conditions of prewar Europe, remarking that:

> We are living in a world in which nothing is made to man's measure; there exists a monstrous discrepancy between man's body, man's mind and the things which at the present time constitute the elements of human existence; everything is in disequilibrium.[10]

Weil is convinced that in her time, science and technology possess collective influence that holds humans and nature captive.[11] Although technology was invented to enhance human life, now the order is most definitely reversed. Human work and nature serve as the means toward the ends of progress, production, wealth and war. The power of nature has been replaced by social power, supported by weapons, machines, and technical knowledge. She notes that the preparation for war spurs "rapid consumption of raw materials and capital equipment, [and] a crazy destruction of wealth of all kinds that previous generations have bequeathed us."[12] Weil fears that the frenzy of militarism and destruction will exceed the limits of social and economic functioning and that European civilization may eventually self-implode. The only limit on the expansion of power lies in its overextension, when a political regime collapses. Then, coun-

---

[10] Simone Weil, *Oppression and Liberty*, trans. Arthur Wills and John Petrie (Boston, MA: University of Massachusetts Press, 1958), p. 108.

[11] For exploration of similarities between Weil and the later Heidegger, see Henry Leroy Fitch, *Simone Weil and the Intellect of Grace*, ed. Martin Andic (New York, NJ: Continuum, 1999), p. 90.

[12] Weil, *Oppression and Liberty*, p. 117.

tries are ruined, populations destroyed, the environment ravaged, all for increase in power to the point of self-destruction. Reflecting on Hitler in 1943, Weil concludes that he displays the flaws of Western societies pushed to an extreme, the supremacy of force enabled by modern science.[13] In many ways, the end of World War II proved her correct.

Feminist thinkers are troubled by associations between power and violence, and consider the patriarchal framework of this association. In her 1939 essay on Homer's *Iliad*, which she refers to as "*le poème de la force*," Weil closely considers the dynamics of social force or power. For Weil, force is grounded in violence, displayed in the Greek and Roman empires as well as modern Europe. Force is addictive, and those who possess it are under the delusion that it can be controlled and held in permanence. There is an immoderation of force that is almost impossible to resist except with 'superhuman' virtue.[14] The Trojan War presages World War II with its massacres, mutilated bodies, enslaved people, dying children, and the contempt for the enemy held by the Greeks, whom she sees not as heroic, but as brutal. Weil observes that gender and social class affect the dynamics of suffering and create disparate levels of victimization.

Weil is particularly concerned with suffering and the body, a theme that pervades her writings on physical labor and selfhood.[15] Counter to Descartes' *cogito ergo sum*, Weil holds that knowledge arises from human experience of forces in the interaction between the body and world. She considers manual labor as the key to understanding human relations to nature. In analyzing factory work, Weil seeks an ideal of liberated labor where there is a proprietary feeling toward the process and product, where attention is unified, and satisfaction is found in the work accomplished.[16] However, an understanding of science and its practical applications is required to circumvent the destructive potential of machine labor. Weil envisions a society where science education allows people to understand technology and its constraints and thus invest work with dignity. It is education that can subordinate technology to the primary values of human flourishing and social justice.

---

[13] Simone Weil, *The Need for Roots*, trans. Arthur Wills (Boston, MA: Beacon, 1952), p. 240.

[14] James P. Holoka, ed. and trans., *Simone Weil's The Iliad or the Poem of Force: A Critical Edition* (New York, NJ: Peter Lang, 2003), p. 5.

[15] For discussion of labor, necessity, and affliction, see Simone Weil, *Notebooks*, trans. Arthur Wills (London: Routledge and Kegan Paul, 1956), pp. 400-404, p. 496.

[16] Weil, *The Need for Roots*, p. 95.

Counter to the evidence of war and relentless force, Weil asserts that the rule of force can be suspended by means of charity, which recognizes others as centers of consciousness as I myself am. In such recognition, I renounce power over others. To do so is what Weil terms "superhuman," which means to manifest grace as opposed to gravity, and to realize God's presence in creation.[17] Although what is superhuman is counter to force, it is not separate from matter and embodiment; rather, it reflects ethical ideals and genuine human existence. According to Weil, what is divine can be found in nature in "the balance or order that is beauty, responsive social relations, [and] methodical work."[18]
The natural world has supreme importance in cognition and perception of God. God is found – in pain, in work, in love – through direct bodily contact with physical reality. Weil employs physical metaphors to illustrate how persons can become a point of contact between God and creation. Any person can serve as the medium for God to touch creation, providing consent is given. Such is the significance of the embodied person in relation to God.[19] Yet such consent involves destruction of the "I". For Weil, the self is the power to say "I"; it is the assertion of the self. In order for a human being to become a point of contact with God, the self must disappear. The self can be annihilated, overwhelmed by divine reality. Self-annihilation can occur in two ways: internally, when a person empties the self in selfless love, or externally, by means of affliction from physical and psychological pain.[20] The cross represents both extremes simultaneously. Both ways are very difficult, and more often it is affliction that affects self-annihilation. However, affliction alone has a destructive effect that removes the "I" by external force; "there is nothing worse," remarks Weil in her *Notebooks*.[21] On the other hand, the destruction of the "I" from within involves giving up the positive use of the will for self assertion, and negative use of the will to maintain attention. Weil names this process of voluntary relinquishment of the self 'decreation' where a person becomes transparent in relation to God.[22]

---

[17] Simone Weil, *Waiting for God*, trans. Emma Craufurd (New York: Harper, 1951), p. 151.
[18] Andrea Nye, *Philosophia: The Thought of Rosa Luxemburg, Simone Weil, and Hannah Arendt* (New York, NJ: Routledge, 1994), p. 104.
[19] Weil, *Notebooks*, p. 401.
[20] Simone Weil, *Gravity and Grace*, trans. Emma Craufurd (New York, NJ: Routledge, 1952), p. 24.
[21] Weil, *Notebooks*, p. 337.
[22] Miklos Veto, Joan Dargan, trans. *The Religious Metaphysics of Simon Weil* (Albany, NJ: SUNY, 1994), p. 27.

Self-loss may seem to condone the self-effacement that women undergo in patriarchal societies, whether through objectification or exploitation. However, the relinquishment of the self is a product of Weil's materialist approach to human being and her ethic of creation. Her viewpoint offers opportunities for freedom from patriarchal norms, self-concern, and motives of gain. By no means does a self-less person give up labor or involvement with others. Rather, negation of the self enables fullest engagement in reality and intense social participation, and actions are performed with attention that connects the individual to other persons and the natural world.[23] Weil's integration of self with nature and the divine repairs the dualism between nature and humanity, and offers an embodied notion of agency and God's presence, while her critique of technology and force remains applicable to the global situation today.

MELISSA RAPHAEL

It is notable that until 2003, there existed no booklength Jewish feminist response to mainstream male Holocaust theology, or to women's Holocaust testimonies. In the pioneering book, *The Female Face of God in Auschwitz*, Raphael makes a bold step in drawing on experiences of female concentration camp survivors to show God's presence in Auschwitz.[24] Her approach is controversial and she is keenly aware of potential objections. By using women's experiences to illustrate theological claims, her book may seem to colonize these Holocaust voices, misrepresent their religious intentions, and silence their witness. In the introduction, Raphael specifically addresses Lawrence Langer's accusation that feminist scholarship may idealize women survivors and create a mythology of comparative endurance that falsifies history.[25] She notes that Langer's work as a whole takes a determined anti-redemptive position in relation to Holocaust testimonies, and accentuates accounts of despair and degradation. In reply, she argues that it does not matter statistically how many individuals maintained moral integrity or showed signs of resistance because even a few examples are significant. Unlike a historian who aims to present a representative picture of

---

[23] For discussion of the mystical meaning of work, and connections between Weil and liberation theology, see Alexander Nava, *The Mystical and Prophetic Thought of Simone Weil and Gustavo Gutierrez* (Albany, NJ: SUNY, 2001), pp. 41-44.

[24] Melissa Raphael, *The Female Face of God in Auschwitz: A Jewish Feminist Theology of the Holocaust* (London: Routledge, 2003).

[25] Langer, "Gendered Suffering?", in *Women in the Holocaust*, p. 362.

survivor's experiences, her theological project has the goal of interpreting the Holocaust within a Jewish narrative of redemption. In women's Holocaust testimonies, Raphael looks for patterns and signs of God's nearness "read theologically as *midrashim* or narrative commentaries on the presence or face of God in Auschwitz."[26] Examples of caring and resistance among women in the camps are viewed as an anticipation of *tikkun olam*, the mending of the world and repairing of the relation to God.

Raphael's reflection on Holocaust testimony focuses on a core of five published memoirs by women deported to Auschwitz between January and December 1944: Sara Nomberg-Przytyk, Olga Lengyel, Isabella Leitner, Giuliana Tedeschi, and Bertha Ferderber-Salz.[27] Not all are Jewish, and not all are religious. But Raphael makes it her task to discern indications of God's presence and connections to Jewish religious observance. These women's experiences provide illustrations for theological insights. In their testimonies, she finds examples of small yet heroic acts of care, compassion and loyalty. Under the tremendous pressure of Holocaust torture, care for the personhood of self and others, established sacred space and time, and rendered a fragment of Auschwitz holy, a place appointed by God.[28]

Raphael considers the ability to sustain supportive relationships as inherently indicative of divine presence. The meaning of such everyday activities for women is anchored in the gendered practices of Jewish orthodox life: men's commitment to study and male group activities, and women's duties including maintenance of domestic purity. While the masculine production of holiness in the concentration camps involved memorized texts and communal rituals, women maintained spiritual care of (lost) home and family through humane acts. Consistent with her focus on women, Raphael looks for evidence of mutual support in the camps, which are the ethical and practical aspects of Jewish life. These day by day acts of care all count as the "practical restoration of persons as created images of the divine wherever and to whatever degree that obtained."[29] Women mediated God's presence by the washing and cleaning of bodies and objects even amidst filth. Bertha Ferderber-Salz recalls that upon transfer from

---

[26] Raphael, *The Female Face of God in Auschwitz*, p. 9.
[27] Raphael, *The Female Face of God in Auschwitz*, p. 13.
[28] Raphael, *The Female Face of God in Auschwitz*, p. 59.
[29] Raphael, *The Female Face of God in Auschwitz*, p. 71. She draws on the relational thinking of Martin Buber and Emmanuel Levinas who conceive of Jewish religiosity as centered on relation to the other.

Auschwitz to Bergen-Belsen, she, her niece, her sister-in-law, and another young girl shared food, helped wash each other's hair, and sewed each other's garments with needles made from splinters of wood.[30] For Raphael, Jewish observance is defined by actions that display God's intention for the world, namely, to repair the damaged humanity of the other and to reveal the holiness of God. Women in the camps were able to sanctify the body, despite desecration, and to retain relationships of care, despite deprivation.

Parallel to her emphasis on women's role as bearers of God's image, Raphael accentuates female aspects of God. She draws upon the notion of God's Shekhinah, the presence of God with Israel in exile, as a resource for feminist theology. In Jewish mysticism of the *Kabbalah*, the *Shekhinah* is understood as a female emanation of God which represents God's dwelling on earth. The *Shekhinah* mediates between the abstract unity of God and the damaged creation which lies below, waiting to be restored. In *kabbalistic* thought, the *Shekhinah* bears maternal qualities of dwelling and protecting. Raphael states: "As *Shekhinah*, God suffers the conditions of finitude. But as God, she endures forever."[31] While Israel, the bearer of God's creative love, lay wounded in the Holocaustal mire, God too suffered in Auschwitz. Employing theological aesthetics, Raphael locates God in the powerless and beautiful face of the female other, immanent in suffering.

Protest and refusal to accept traditional theodicy explanations are common features of Holocaust theology. However, Raphael does not accuse God. She is critical of the work of major Jewish theologians such as Richard Rubenstein, Eliezer Berkovits, and David Blumenthal, who are troubled by the indifference of the almighty God to Holocaust suffering. She observes that their protests against God presume a patriarchal model of a sovereign ruler who is inscrutably absent, silent, impotent, or cruel. In fact, these male authors accuse God of not being patriarchal enough, otherwise he would have prevented the Holocaust. They operate with masculine norms in their assumptions about God's interaction with humanity as a powerful overlord. Making a surprising comparison between God and Nazi authority, she observes that Emil Fackenheim's 'commanding voice' of God insisting that Jews, remain Jews is similar in form and type to Hitler's coercive and oppressive commands. Not only is a commanding God masculine, but biblically, God's commandments at Mount Sinai were delivered

---

[30] Raphael, *The Female Face of God in Auschwitz*, p. 78.
[31] Raphael, *The Female Face of God in Auschwitz*, p. 125.

only to men who, for the sake of purity, had kept themselves apart from women. Women are distanced from divine revelation in history, and alienated from the patriarchal God whose sovereign power is "too close in kind to the world domination sought by Nazi Germany."[32] In reflecting on violent power, Raphael is uncomfortable with how the modern state of Israel may be viewed redemptively in post-Holocaust thought. Since Jewish propriety over the land requires sustained military action, it validates domination as a sign of divine favor. Human use of force to accomplish God's aims condones violence against nature, the land of Palestine, and its inhabitants.[33]

By placing priority on embodiment and relation, Raphael understands the people of Israel as God's visible presence in the world, and God's power as manifest in relational care. The mystical notion of the *Shekhinah*, God's presence with Israel, represents the vulnerability of God in relation. Addressing the Holocaust and nature, Raphael's response affirms God's manifestation in bodily dwelling.

CONCLUDING THOUGHTS

Bringing together these two female thinkers from different eras creates a context for dialogue on numerous fronts. Indeed there is potential for a larger comparative project for a futher examination of these authors, in the contexts of Holocaust and feminist scholarship. This chapter cannot do justice to the many issues raised. At this point, I shall conclude with a few remarks about major themes and their significance. Both Weil and Raphael expose harmful attitudes towards nature involving misconceptions and misuse of power. They critically examine notions of force represented in history by male agents, reinforced by technology, and used with horrific devastation in wartime. Weil's account of power and justice provides an insightful description of conditions in contemporary society. Raphael imagines an alternative to dominating power, using female experiences and accentuating human and divine vulnerability. Weil also moves in this direction with her reflections on loss of the 'I' and openness to God. They agree that notions of human power, and God's power, are inappropriately portrayed as forceful.

In contrast with contemporary feminist perspectives, Weil's approach to embodiment could pay more attention to the specificity of gender in considering

---

[32] Raphael, *The Female Face of God in Auschwitz*, p. 36.
[33] Raphael, *The Female Face of God in Auschwitz*, pp. 31-32.

power and self-loss. Her autobiographical reflection provides subtle perspectives on female identity; however, her philosophy of religion often operates with an ungendered self. On the other hand, Raphael risks dualizing men's and women's experiences with her conclusions about the female face of God present in women's caring. In so doing, she lies open to the accusation that she glorifies women over men. But in my reading, she does not actually deny God's presence among men, men's relational abilities, or men's theological contributions. She is clearly appreciative of the male kabbalistic tradition, and the insights of Jewish philosophers such as Emmanuel Levinas and Martin Buber, who accentuate the ethical and religious significance of interpersonal relation. For myself, as a feminist who prioritizes the incorporation of male and female participation in constructive proposals, I would hope to see Raphael's conclusions extended more broadly to include men in an ethics of care that manifests the restoration of creation in *tikkun*.[34]

In contrast with much previous theological reflections on the Holocaust dwelling on theodicy, it is striking that protest is not enacted by these authors. Instead, their approaches are mystical, in the sense that they rest on direct transformative manifestation of God's presence. Raphael develops feminist implications of God's Shekhinah dwelling with Israel, while Weil employs a negative theology found among prominent Catholic mystics. Suffering is important to both thinkers who find God in situations of created vulnerability. They offer ways to imagine God's nearness in Holocaust times and in the present. The classical conceptual separation of man, nature, and God – where God holds power over man, man has dominion over nature, and man protests God's actions or neglects – is manifest in the alienation displayed both in Holocaust times and in our technological age. Weil and Raphael share common intuitions that reflect wider impulses in contemporary feminist reflection, as well as post-Holocaust theology and ethics. Their proposals are open to critique and revision. But ultimately, it is admirable that Weil and Raphael undertake the crucially important, and immensely complicated task of developing a holistic vision of nature, humanity, and God.

---

[34] Scholars have found profound examples of men's care and mutual help; for example, see Nechama Tec, *Resilience and Courage: Women, Men, and the Holocaust* (New Haven, CT: Yale University Press, 2003), pp. 188-191.

In Response to Sarah Pinnock

Didier Pollefeyt

For Sarah K. Pinnock, the Holocaust is the outcome of modernity, and more specifically modern technology, which she associates with manipulation and domination of nature, typical for a male approach of reality. She proposes to overcome the dualism that is at the basis of the devaluation of nature (female) by technology (male) by including women's voices and feminist perspectives in Holocaust reflections to become aware of the realization of God's presence in nature, as it was even experienced *in* the Holocaust by (female) victims. The theological idea is to rediscover the divinity of nature ('divine immanence') as a source of resistance against violence, war and genocide. Pinnock's analysis is in line with an important body of Holocaust literature that has demonstrated in a convincing way the relation between the Holocaust and modernity, especially modern technology[1]. The originality of her contribution is to be found in the connection between technology and gender dichotomies reaching a culmination point in the Holocaust. As a consequence, she pleads for the rediscovery of 'God' or the 'divine presence' in direct bodily contact with the psychical reality.

But what if we just turn upside down this analysis? What if the Holocaust is not just an outcome but rather a misuse, a perversion of modern technology and what if Pinnock's theological conclusions, especially the concepts of God and nature she uses, seem to have more resemblances with the nazi concepts of God and nature, then is clear on first sight? I am wondering how a feminist reading on Holocaust and nature would deal with these opposite presuppositions to understand nazi ideology which also have a ground in established Holocaust studies.

In her chapter, Pinnock refers with different words to the divine reality: 'God', 'the female of God', the 'wholeness of God', 'the divine', a 'female emanation of God', 'divine immanence', etc. Is it correct to say that she refers to an immanent 'God', an anonymous power, God as Mother Nature, etc.? In this way, Sarah Pinnock comes close to the panentheistic God as 'He' is understood 'after Auschwitz' by Richard Rubenstein. In confrontation with the Holocaust, Rubenstein rejected radically the traditional understanding of the transcendent God father and opted for an immanent concept of God, a God beyond God: "God is

[1] Zygmunt Bauman, *Modernity and the Holocaust* (Cambridge: Polity Press, 1989).

the ocean and we are the waves. In some sense each wave has its moment in which it is distinguishable as a somewhat separate entity. Nevertheless, no wave is entirely distinct from the ocean which is its substantial ground"[2]. As Pinnock, Rubenstein speaks of the divine using female terminology: the abyss, the source, the ground, even: the sacred womb. But Rubenstein's God is not the warm, friendly, gentle God as we know him in a lot of contemporary New Age literature. Rubenstein's theology could be called the radical, dark – or realistic – side of a lot of popular contemporary New Age movements. His God is shockingly not a very loving God. God as 'Holy Nothingness' (Rubenstein) is an anonymous power, without moral concern, beyond every human understanding. His presence is dangerous and can destroy us in just one second. She is not a loving Mother, but a cannibalistic Mother that devours the fruits of her own womb and brings salvation by killing. My first question to Sara Pinnock is how her concept of God relates to the post Holocaust view on God of Richard Rubenstein.

It is well-known that the Jewish philosopher Emil Fackenheim has accused Rubenstein's theology as a 'posthumous victory for Hitler'[3]. In fact, by declaring the traditional God of history death and by returning to the God of nature, Rubenstein is doing, in the view of Fackenheim, what Hitler did not succeed in doing, bringing Hitler's work to an end. Fackenheim's critique also forms a challenge to the reflections of Pinnock. The question is then: does the post-Holocaust feminist reflection on nature not bring an identification of God and nature which is very close to the nazi concept of God, namely a purely immanent God, a violent, anonymous power, a merciless reality that kills first the weakest, a concept of the divine related to *Blut und Boden*, etc. Did not exactly the Nazis divinize the powers of nature and deduced laws from it that legitimized their crimes? How can we ever trust Nature with a capital 'N'?

In the Jewish tradition, and also in the thinking of Emil Fackenheim, God reveals himself not in nature, but as a resistance against nature, as a transcendent commandment not to grant Hitler posthumous victories. And for that other Jewish philosopher, Emmanuel Levinas, nature is considered as the *il y a* ('there is'), the anonymity, the reality that does not respect the face of the other. The absolute

---

[2] Richard Rubenstein, *Morality and Eros* (New York, NJ: McGraw-Hill, 1970), p. 186.

[3] Emil Fackenheim, "The 614th Commandment", in *Judaism* 16(1967)269-273, p. 271: 'The authentic Jew has the duty not to hand Hitler posthumous victories'.

does not come from nature (*being*) but from beyond nature (*otherwise then being*), for the victims of the struggle of life inherent to the dynamics of nature[4].
This brings me to Pinnock's critique on the role of technology in the origin of the Holocaust. It is true that modern technology played an immense role in creating the possibility of the Holocaust, not only thanks to concrete technologies, but also, and especially because of the 'technological mentality' of the modern nazi man. Nevertheless, one can argue that the technological mentality is wrongly presented by Pinnock as the *motivation* to exterminate so many people, while it's role was in the first plays restricted to *facilitate* its execution. Technology did help the extermination, but it was not its reason. Even if the technological attitude was one of the preconditions for the process of extermination, it cannot explain the Holocaust as such. In other words, it is not because the Holocaust used technology that a technological world forms as such a Holocaust. E.g., it is not because Eichmann was a bureaucrat that every bureaucrat is a mass murderer. Technology can only exercise its destructive power, when a political ideology takes the overhand that legitimizes this power. In my view, this political ideology has much deeper roots in pre- and anti-modern pagan philosophies then in the modern world view. In fact, Hitler tried to defeat modernity with its own means by using selectively some modern elements. Modernity and even modern technology cannot be understood without its universal ethos that created human rights. The same holds true for modern technology which was created to humanize human work, to heal the sick, to create more leisure time for more people, etc. In this sense, the nazi use of modern technology is more a *misuse*, a perversion, then the revelation of it's own essence.
Modernity remains a very ambivalent reality. The same holds true for technology. And also for the concept of nature and of God. I agree on this with Sarah Pinnock. In her chapter, Pinnock asks to transcend dualistic categories, but my last question is if her own analysis remains not imprisoned in dualistic categories itself: female *against* male, nature *against* power, body *against* mind, immancence *against* transcendence, sensitivity *against* technology, etc. Should a feminist thinking after Auschwitz not be more critical against all categories, not

---

4 Didier Pollefeyt, "The Trauma of the Holocaust as a Central Challenge of Levinas' Ethical and Theological Thought", in Yehuda Bauer, *Remembering for the Future II* (Papers to be presented at an international scholars' conference to be held in Berlin, 13-17 May, 1994) *Theology I: Theological Culture of Remembrance and Ethics* (Oxford/New York, NJ: Pergamon Press, 1995); Emmanuel Levinas, *Difficile liberté. Essais sur le judaïsme* (Paris: Albin Michel, 2° ed., 1976) p. 201.

only the categories it criticizes traditionally and with good right, such as male, power, transcendence, technology, etc., but also *vis-à-vis* its own categories so easily taken for granted, questioning them when encountering the Holocaust: female, nature, body, immancence and sensitivity.

In Response to Sarah Pinnock

Rochelle L. Millen

"Veshakhanti betocham: And I will dwell among Them"

The verse above from Exodus 25:8 relates to the indwelling attribute of God in Hebrew Scriptures and is identified with the grammatically feminine name of God, *Shekhinah* (from the root 'to dwell within'). The womb, that which nurtures and protects, that which is within and is integral to life: all these are aspects of God regarded as feminine, based on the physiological processes associated with conception, pregnancy, and birth. God is not only Father and King, but also Mother and Source of Life.

In her essay, *Feminist Reflection on the Holocaust and Nature*, Sarah Pinnock analyzes the perspectives of two Jewish women philosophers as they relate to this overall theme. The connection to 'nature,' however, is sometimes elusive. Pinnock begins by referring to the Aristotelian distinction between natural and moral evil, later appropriated by both Maimonides and Aquinas. Her conclusion is that the line drawn between these two types of evil in Aristotelian theory has been blurred by developments in modern science – a debatable premise. Then she concludes by saying "It is troubling that the divide between what is moral and natural potentially makes nature seem irrelevant to Holocaust reflection." The rest of Pinnock's essay is meant to demonstrate that nature is not only relevant, but that the relation to the Holocaust also can be illuminated from a feminist viewpoint. Despite the fact that the thinkers discussed are both fascinating female – and Raphael is a declared feminist – the essay only hints at possible ways in which their writings manifest and illuminate the intertwining of the Shoah and Nature from a feminist perspective; there is not a clear connection. The meaning of 'nature' shifts from Aristotelian natural evil to mother earth to temperament to the embodiment of spirit. That each represents a feminist reflection, perhaps emphasized by the authors' femaleness, is possible, but uncertain. And only in the case of Raphael does the feminist stance have a direct theological result. Pinnock states that she focuses on "female authors [...] to include more women's voices in dialogue and not to create gender dichotomies [...] women's experiences and feminist perspectives are broadly relevant to Holocaust reflection." The works of Simone Weil and Melissa Raphael, and Pinnock's discussion of them, offer some insights into post-Holocaust theology, but whether they are 'feminist' and closely connected thematically to the Holocaust and Nature may be questioned, especially in the case of Simone Weil.

Weil, born to an assimilated Jewish Parisian family, became a Christian neo-Platonist. The decision to convert to Roman Catholicism was deterred by her disdain for Hebrew Scriptures and its allegedly unfortunate influence on the formation of Christian doctrine. The twelfth century Albigensians, the very same massacred (beginning in 1209) due to the edict of Innocent III, represent to her the purest form of Christianity, and the Gnosticism she embraces, resonates with ideas also found in the *Lurianic Kabbalah*. Weil thought that Hebrew Scriptures manifested a "fundamental error concerning God,"[1] that anything spiritually correct in Judaism came about due to the exile, i.e. to the cultural influences of Persia, Greece, and Chaldea. She sees Judaism as inseparable from idolatry due to the concept of chosenness and what she understands as an inherent nationalism. Christianity elevates spirituality not only through Jesus on the cross, but also through the interpretations of the early church fathers, especially Marcion and Tertullian. The emergence of the Christian God is a sign of great spiritual evolution and progress over Jehovah of the Hebrew Bible. For Weil, God is love, and "Before all things, God loves himself [sic]."[2] Creation is not the exercise of divine power, but rather the abdication of God, even the sacrifice of God. Why God abdicated God's power in favor of cosmic necessity, remains a mystery to Weil. The human task is to consent to necessity, and suffering in the world, since God so consented (both in creation and in Jesus on the cross). And it is by consenting to God's will that humans can atone for the sin of our existence and become nothing. For the self – in all its aspects – is illusory. It is governed by necessity, and therefore a love of suffering, which Weil terms 'affliction' in its most intense state, is the highest level of the love of God.. God can never be perfectly present to us in the material world, since we are flesh. But God can be perfectly absent from us in extreme affliction, and affliction –the cross – is at the very center of Christianity. The erasure of the 'I' through affliction is what makes possible God's grace.

Weil defends God's justice while destroying the possibility of human justice. She has formulated a negative theodicy in the figure of the crucified God. In a

[1] Weil, *Letter to a Priest*, p. 106.
[2] Weil, *Waiting on God*, p. 85.

sense, her radical theology has found a theological use for evil; she passionately, desperately, even brilliantly seeks to justify and rationalize human suffering.
The social thought of Weil, as that of Buber and others in mid-twentieth century, rails against the collective, and in this, indirect fashion is concerned with the Holocaust.. But at the same time there is a masochism undergirding that which the individual should search out. She is writing about Nature insofar as her philosophical thinking must contend with matter, and the 'love of affliction' resonates with the palpable spirituality of the medieval Christian mystics.
Ascribing the negative elements in Christian thought to 'Hebrew prejudices,' it does seem tasteless, as a friend of mine commented, for Weil to have preached the gospel of affliction from England in 1942, even while starving herself to death. It would have been more authentic to travel to Poland and announce that she was Jewish. Living in the time of Hitler and Stalin, Weil's Christianity was like that of the early Christians, who expected comfort only from God's hovering presence. In contrast to Weil, who is a pre-modern feminist[3] and does not address the Holocaust directly, Melissa Raphael writes from the perspective of a well-versed theologian, not only in Holocaust literature and theology, but also in contemporary feminist texts. Raphael's central concepts, well-summarized by Pinnock, focus on God as *Shekhinah*, the indwelling Presence of Mother-God. Through the humanity of gendered acts, such as washing, recalling recipes, becoming a pseudo-mother or daughter or sister or aunt, Raphael understands God in God's female attributes to have been present in Auschwitz. Only in the male sense of God as powerful, authoritative, and commanding, God was absent or hidden or even abusive. Raphael uses the famous *midrash* of God accompanying the Jewish people in exile and sharing their suffering[4] as a means of understanding God's nearness, even in the midst of the concentration camp universe.
Raphael writes: "Patriarchy has ranked and used up creation [...] God has been exiled by the violent refusal of her peace [...]. If Auschwitz meant that God could no longer behold God on earth, then her being and her reason for creation was close to destruction. This was the case when God seemed so dispersed by absolute atrocity to have disappeared altogether." She continues: "In those moments when God could behold God in the midst of Auschwitz [...] these were moments

[3] She lived after the suffragettes of the late nineteenth and early twentieth centuries, but prior to modern feminism.
[4] See BT Megillah 29a.

of *tikkun* – the restoration of God's first vision [...] of the world."[5] Raphael ascribes those acts of restoration (almost in a kabbalistic sense) primarily to women, for whom relationship, cleanliness, nurture, and protection manifest aspects of holiness, and therefore of God's presence, amidst the horror. Raphael asserts that God's alienation from the world is a direct outcome of the patriarchal view of the Divine, predominant for centuries.

Feminist theology, both Christian and Jewish, have distilled and struggled with the ramifications of this notion.[6] Yet despite the astuteness of sections of Raphael's work, that female acts of hesed, or lovingkindness brought God to Birkenau or Ravensbruck, remains a difficult idea. Why not similar acts of men? Why not the prayers of men – and women, or the trial of God told to us by Elie Wiesel? What about the Hasidic tales told by Yaffa Eliach?[7]

For Simone Weil, God cannot be the *Shekhinah* , as God's grace and love are the result of affliction, decreation, and suffering. For Melissa Raphael, the Holocaust remains "theologically instructive,"[8] in that "there is no actuality that is not always transfigured by an aesthetic sense of the presence of God and its moral demand. The face of the other as a perceptible image of God is always before us."[9] This is Buberian; it is Levinasian.; and it acknowledges the possible presence of the *Shekhinah* even in Sorbibor and Treblinka. Yet to affirm the Source of Life hovering amidst the horror, present in the raspberry given to Gerda Klein or the conversations of Ruth Kluger or Margot and Franci, mother of Helen Epstein[10], in Theresienstadt, remains a challenge. I cannot fathom Mother God in such places any more than God the Father. Intriguing, engaging, and theologically enlightening, the writings of Weil and Raphael shed a shadowed light upon the theme of the Holocaust and Nature.

---

[5] Melissa Raphael, *The Female Face of God in Auschwitz: A Jewish Feminist Theology of the Holocaust* (London and New York, NJ: Routledge, 2003), pp. 158-159.

[6] Examples of this extensive literature are Judith Plaskow, *Standing Again at Sinai*: *Judaism from a Feminist Perspective* (San Francisco, CA: Harper and Row, 1990); Rachel Adler, *Engendering Judaism: An Inclusive Theology and Ethics* (Boston, MA: Beacon Press, 1999); Rosemary Ruether, *Sexism and God-Talk: An Inclusive Theology* (Boston, Ma: Beacon Press, 1993).

[7] Yaffa Eliach, *Hasidic Tales of the Holocaust* (New York, NJ: Oxford University Press, 1982).

[8] Raphael, p. 17.

[9] Raphael, p. 18.

[10] Helen Epstein, *Where she Came From: A Daughter's Search for her Mother's History* (New York, NJ: Penguin, 1998).

In Reply to My Respondents

Sarah Pinnock

I appreciate the probing questions posed by Rochelle Millen and Didier Pollefeyt. Their comments accentuate the challenges I face in bringing feminism, the Holocaust, and nature into dialogue. My response aims to clarify my assumptions and the framework of my approach in order to address their concerns.

Ontological connections between the subjects of nature and women are inescapable from a feminist standpoint. Both women and nature are viewed as passive and subordinate to male activity in classical philosophies, and patriarchal attitudes toward women and nature translate into acts of violence that entrench superiority and oppression. Such hierarchical dualisms are also found in Nazi ideology about women's domestic and reproductive roles serving the German nation. Evidently, these dualisms need to be challenged. My approach seeks the revaluation of the subordinate member of the binary, which is not intended to entrench static poles. Rather, the possibility of overcoming false oppositions depends on working through the pejorative dualisms themselves.

While the association between women and nature is a product of patriarchal thinking, it can also work constructively to advance new perspectives. My strategy in choosing female thinkers is to consider how they respond to patriarchal strictures, nature, and the Holocaust both implicitly and explicitly. Millen and Pollefeyt raise important methodological questions about my affirmative prioritization of nature. I agree with Pollefeyt that nature should not be exempt from critique, although I highlight the thematic affinity between women and nature in my essay. I also appreciate Millen's observation that there is more discussion of nature than the Holocaust in my essay. There is an asymmetry in my consideration of the Holocaust using a feminist perspective on nature.

My approach to the Holocaust considers moral and natural evil, and both human and divine responsibility. Classical theodicy defines natural evil as the destruction wrought by natural forces such as storms or earthquakes. Although the link between women and nature implies that both are passive with respect to dominating forces, they also have the capacity to be agents causing harm. Just as nature can be destructive, the capacity of women for evil is not mitigated by the fact of sexism. Increasingly, military technology enlarges the human capacity to destroy and deployment does not depend on bodily strength.

Technology has the positive ability to alleviate nature's deadly force, but economic and political factors are paramount in determining whether technology helps. Millen expresses skepticism about whether advances in technology blur boundaries between moral and natural evil, as I claim. My point is that there myriad opportunities to alleviate natural disasters and quite often there is not the political or economic will to do so. It is ethically culpable that governments and businesses are so unsuccessful in prevention and rescue from natural evils. The 2010 Gulf of Mexico oil spill illustrates the need for more moral responsibility on the part of corporations to protect nature. In the same Gulf region, the impact of Hurricane Katrina in 2005 shows how natural evil blurs with moral evil when failures in flood management and aid to victims created a much larger disaster. Pollefeyt is correct that in the Holocaust, ideology rather than technology was the motivation for Nazi killing. Nevertheless, Nazi violence displays the enormous power of technology to exploit nature and human bodies. Along with authors such as Martin Buber and Simone Weil, I believe that technology creates an instrumental relationship that reduces everything to an object to be used, rather than an ethical subject. The Holocaust discloses the evil implications of this instrumentalism most vividly.

Simone Weil and Melissa Raphael are thinkers who develop connections between God and suffering, which engage the Holocaust. Millen points out salient differences between the two authors and various objections to their ideas. My reason for pairing these authors was to draw out points of contact in their conceptions of God and nature and the search for God in suffering and evil, rather than focus on criticism of their proposals. Quite deliberately, I chose to explore their appreciation of nature in constructive directions and to examine how the Holocaust poses theological questions.

Millen points out Weil's alienation from Judaism and criticizes her mysticism. It is true that Weil rejected her Jewish heritage and embraced secular French identity. Weil is dismissive of the Hebrew Bible while her philosophical interests led her towards ancient Greece and India and medieval Christian mystics. However, she did not take an indifferent attitude towards National Socialism and the Holocaust. Weil worked for the French Resistance and wrote articles condemning Hitler. In 1943, her parents accompanied her to the United States to escape the Final Solution, but she returned to Europe to join the war effort in London on behalf of the French government in exile. Her religious reflection on suffering led her to seek God in affliction and her poor health was exacerbated by her unwillingness to eat anything in excess of meager war rations.

Like Weil, Raphael also finds God in affliction. She studies memoirs of concentration camp survivors and discovers the divine face of God in women's experiences of care. Raphael describes the indwelling of God as the presence of the *Shekhinah* found among both Jewish and non-Jewish women. Although her theological interpretation is Jewish, Raphael believes that God can be seen in the lives of people who are non-religious. God's presence is ethical and relational. In comparing these authors, Raphael's direct engagement with survivor literature helps connect Weil's mystical ideas about God more closely to the Holocaust. Both of them locate God alongside suffering rather than as a figure that stands above history.

Elie Wiesel and many Jewish and Christian theologians object to the notion of an omnipotent God in the context of genocide. They find it morally repugnant to assert that God plans the Holocaust for redemptive purposes. Furthermore, feminist thinkers observe that an all-powerful God correlates with authoritarian religious and political institutions. There is a near consensus among feminist theologians in affirming an alternative concept of God centered on love that includes the passivity of suffering. As Weil and Raphael suggest, such a God can be found even in Holocaust situations of degradation. Millen remarks with honesty that: "I cannot fathom Mother God in such places any more than God the Father." She does not agree with Raphael and Weil who situate God's presence in severe suffering. Rather, God is found in other places. For instance, survivors bear witness to Jewish prayers and rituals for holy days performed in the camps, and Hasidic tales depict escape attributed to divine protection. I concur. Nevertheless, I think that the focus is on nature makes suffering particularly appropriate.

In important ways, the passivity of human victims is analogous to the passivity of nature. Yet human beings are the agents of the Holocaust whereas nature is entirely dependent. The Jewish mysticism of the Kabbalah portrays creation as fractured with shards of divine light scattered throughout creation. For Weil and Raphael, the Holocaust displays the abuse of nature with mere glimmers of divine immanence. They wrestle with the paradox of God's absence and presence in nature. Weil places more emphasis on emptiness and void, whereas Raphael focuses on seeing God's face in human faces. Both authors consider God as hidden and present physically to persons, irrespective of their participation in religious institutions.

Pollefeyt compares the immanence of God developed by Weil and Raphael to the deity of holy nothingness proposed in *After Auschwitz* by Richard Rubenstein, but this connection is misleading. Rubenstein's God is amoral, whereas the feminist divine is found in charity. Rubenstein's God is destructively powerful, whereas the God described in my essay appears in suffering and weakness. Frankly, I am startled by Pollefeyt's suggestion that feminist religious beliefs about nature are analogous to Nazi pagan religion. Feminist claims about the closeness of God and nature are universal and relational. The views I explore have nothing to do with racism, social Darwinism, or the Germanic myth of blood and soil. It is true that Weil and Raphael do not describe the traditional Jewish God of history and this makes them controversial. But an unorthodox theology after Auschwitz does not translate into a posthumous victory for Hitler.

I wish to conclude with the observation that the Holocaust and nature are typically approached from two historical perspectives: Nazi and Jewish. Since Nazi beliefs were self-consciously constructed in opposition to Judaism, these two standpoints are a foil for one another. Moreover, Jewish perspectives provide critical leverage on the genocidal Nazi affirmation of nature. Thus, a dualism emerges between Nazi nature religion and Jewish views of land and, in particular, Israel. I perceive that Nazi paganism raises grave suspicions against the immanence of the divine in nature, as pointed out by Pollefeyt, which deserves investigation. I also appreciate that the Jewish identity of my female authors and the proximity of suffering to God in their thinking raise complex questions, as Millen indicates. Both Nazi and Jewish standpoints are normative in Holocaust reflection on nature and my approach blurs both boundaries. Difficult questions arise about the immanence of God. Yet I am convinced that it is worth navigating in the midst of the dichotomy between what is pagan and Jewish to affirm the religious appreciation of nature.

"SKULL"

Arie A. Galles

1

December 13, 1993

As I listen to Bach on the tape deck I work on the 'Bergen – Belsen' drawing. 'Konzert für 4 Cymbali und Orchester a-moll', conducted by Karl Richter, with the Münchener Bach-Orchester. I listen and draw to the set of tapes I bought nearly seventeen years ago. I occupy myself exclusively with the dark woods and white roads outside of the camp. After much research and preparatory studies, this is my first full-scale drawing of the *Fourteen Stations/Hey Yud Dalet* suite. The area my charcoal creates allows me to concern myself only with texture and the relative values I am incising upon the paper. The music permeates my head. I enjoy the reverie.

I draw four dots, four tiny black dots on the north west perimeter of the camp. Suddenly, the reality of what those dots are, hits me with the force of a sledge-hammer against my heart. These are shadows of guard towers. When this photograph was taken, the towers were manned by the SS. I can feel the camp overflowing with its tortured prisoners.

I break down in tears, and am unable to draw anything at all. I call Sara and we go for a walk. She understands. For five years after the war, she herself lived in Bergen-Belsen after the soldier's quarters were turned into a Displaced Persons Camp.

2

December 25, 1993

Just completed 'Station # 5, Bergen-Belsen'. This is probably the most competent drawing I have made in my life. Looking at it, I sense an actual depth of space between me and the concentration camp. I can almost breathe the air rising above it. I believe I could stretch out my hand and feel the wind upon my extended fingers. The camp is so very far below.

It is a strange complex of buildings, straggling a highway and cutting it in half with its barbed wire outline. The white spots are ash pits and mass graves. On the center right of the camp, and just outside the fence, a skull grins at me from a clearing. The skull is the clearing. Three clumps of trees make up its eye sockets and nasal cavity, a glimpse of a secondary road shines through the trees. A white toothy macabre smile!

I saw it immediately, as I first set eyes on this old RAF photograph taken on September 13, 1944. The skull is not far from the Men's Camp, right past the latrines. Is Nature screaming to heavens the nature of this place? The original photo bears a notation, perhaps by someone from the *RAF Air Reconnaissance*. It is a small circle enclosing a T junction just to the right of the clearing. Next to the circle, written in white against the dark background is, *X 471664*. I do not include this in the finished work.

People viewing the drawing may ask why I drew a skull there. I can only respond that I didn't invent it, I drew what I saw in front of my eyes.

Personalia

John K. Roth is the Edward J. Sexton Professor Emeritus of Philosophy and the founding Director of the Center for the Study of the Holocaust, Genocide, and Human Rights at Claremont McKenna College, where he taught from 1966 through 2006. In addition to service on the United States Holocaust Memorial Council and on the editorial board for *Holocaust and Genocide Studies*, he has published hundreds of articles and reviews and authored, co-authored, or edited more than forty books, including, most recently, *Genocide and Human Rights: A Philosophical Guide*; *Gray Zones: Ambiguity and Compromise in the Holocaust and Its Aftermath*; and *Ethics during and after the Holocaust: In the Shadow of Birkenau*. Roth has been Visiting Professor of Holocaust studies at the University of Haifa, Israel, and his Holocaust-related research appointments have included a 2001 Koerner Visiting Fellowship at the Oxford Centre for Hebrew and Jewish Studies in England, as well as a 2004-05 appointment as the Ina Levine Invitational Scholar at the Center for Advanced Holocaust Studies, United States Holocaust Memorial Museum, Washington, D.C. In 1988, Roth was named U.S. National Professor of the Year by the Council for Advancement and Support of Education and the Carnegie Foundation for the Advancement of Teaching. During the 2007-2008 academic year, he will be the Robert and Carolyn Frederick Distinguished Visiting Professor of Ethics at DePauw University in Greencastle, Indiana.

Sarah Pinnock is associate professor of contemporary religious thought at Trinity University in San Antonio TX. Her research bridges the fields of philosophy of religion, theology, and ethics. Special areas of interest encompass theodicy, the Holocaust, death and dying, and feminist religious thought. Selected publications include "Mystical Selfhood and Women's Agency: Simone Weil and French Feminist Philosophy" in *The Relevance of the Radical: Simone Weil 100 Years Later* (2010) and "Atrocity and Ambiguity: Recent Developments in Christian Holocaust Responses" in the *Journal of the American Academy of Religion* (2007). She authored the book *Beyond Theodicy: Jewish and Christian Continental Thinkers Respond to the Holocaust* (2002) and edited a volume on feminist political mysticism, *The Theology of Dorothee Sölle* (2003).

Rochelle L. Millen, Ph.D., is Professor of Religion at Wittenberg University. Recipient of many grants and awards, Millen has authored numerous book chapters and essays. She is editor of *New Perspectives on the Holocaust: A Guide for Teachers and Scholars* (NY: New York University Press, 1996), author of *Women, Birth, and Death in Jewish Law and Practice* (Hanover, NH: University Press of New England, January, 2004), and co-editor of *Testimony, Tensions and Tikkun: Reflections on Teaching the Holocaust in Colleges and Universities*(University of Washington Press: 2007). She is co-founder and for seven years served as co-chair of the Religion, Holocaust, and Genocide Group of the American Academy of Religion. Millen serves on the Academic Advisory Board of the Hadassah-Brandeis Institute, is a founding participant of the Weinstein International Holocaust Symposium, a member of the Board of the Ohio Council on Holocaust Education, and on the editorial board of the Stephen S. Weinstein Series in Post-Holocaust Studies of the University of Washington Press. She also serves as a member of the Church Advisory Committee of the USHMM. Millen chairs the Professional Development Committee for the Columbus Jewish Federation, and in1995 received the prestigious Samuel Belkin Memorial Award for Professional Achievement from Stern College for Women of Yeshiva University. Currently, she is working on a volume on interwar Poland using a private correspondence from 1930-1936.

David Patterson holds the Bornblum Chair in Judaic Studies at The University of Memphis. A winner the Koret Jewish Book Award, he has published more than 125 articles and chapters on philosophy, literature, Judaism, and Holocaust Studies. His writings have been anthologized, and his more than two dozen books include *Honey from the Rock: Jewish-Christian Dialogue—The Next Step* (with Alan L. Berger, forthcoming), *Open Wounds: The Crisis of Jewish Thought in the Aftermath of Auschwitz* (2006), *Wrestling with the Angel* (2006), *Hebrew Language and Jewish Thought* (2005), *Along the Edge of Annihilation* (1999), *Sun Turned to Darkness* (1998), *The Greatest Jewish Stories Ever Told* (1997), *When Learned Men Murder* (1996), *Exile* (1995), *Pilgrimage of a Proselyte* (1993), *The Shriek of Silence* (1992), *In Dialogue and Dilemma with Elie Wiesel* (1991), *Literature and Spirit* (1988), *The Affirming Flame* (1988), and *Faith and Philosophy* (1982). He is the editor and translator of *The Complete Black Book of Russian Jewry* (2002) and co-editor (with Alan L. Berger) of the

Encyclopedia of Holocaust Literature (2002), as well as co-editor (with John K. Roth) of *Fire in the Ashes* (2005) and *After-Words* (2004).

Peter Haas received his B.A. in Ancient Near East History from the University of Michigan in 1970 and then attended Hebrew Union College in Cincinnati, from where he received ordination as a Reform rabbi in 1974. After ordination, he served as an active U.S. Army chaplain for three years. Upon completion of active duty, Rabbi Haas enrolled in the graduate program in religion at Brown University, earning a Ph.D. in Jewish Studies in 1980. Joining the faculty at Vanderbilt University in 1980, he taught courses in Judaism, Jewish ethics, the Holocaust, Western religion, and the Middle East Conflict. He joined the faculty of the Department of Religious Studies at Case Western Reserve University in January, 2000, and was appointed chair of the department in 2003. He is also a visiting professor at the Spertus Institute of Jewish Studies in Chicago, IL. Professor Haas has published several books and articles dealing with moral discourse and with Jewish and Christian thought after the Holocaust. He has lectured in the United States, Germany, Italy, Belgium and Israel. His most recent work is on human rights in Judaism.

Arie Galles is Professor of Art and Director of Creative Arts at Soko University of America in Aliso Viejo, California. Apart from his academical career as an instructor and professor at several universities in the USA, Galles is known as an artist. His most famous projects were his *Reflected-Light Paintings and Drawings*, and his charcoal drawings named *Fourteen Stages/Hey Yud Dalet*. During the years he worked als the *Fourteen* Stages, he kept a daily journal, of which parts are adopted in this book. Galles was influenced by his Jewish background; He was born in Uzbekistan in the aftermath of Word War II. He resided in Poland, Israel and Italy. Since 1958 he lives in the USA. His works were shown during several exhibitions, as well as in solo shows at several galleries, including the O.K. Harris Gallery, New York City, and the Zolla Lieberman Gallery, Chicago.

Didier Pollefeyt, full professor in Pastoral Theology, Theology of Jewish-Christian Relations and Post-Holocaust Theology and vice-dean of education at the Faculty of Theology, Katholieke Universiteit Leuven, Belgium. He holds the Chair of the *Center for Teacher Education in Religion* and is Director of the *Centre for Peace Ethics*, Faculty of Theology, K.U.Leuven. He is a member of the *National Catholic Commission for the Relations with Judaism* of Belgium. He is the founder of THOMAS, a catholic internetproject for religious education sponsored by the Belgian Bishops Conference and the Office for Catholic Education Flanders (www.godsdienstonderwijs.be). (Co-)author of Didier Pollefeyt, ed., *Jews and Christians: Rivals or Partners for the Kingdom of God? In Search for an Alternative for the Christian Theology of Substitution* (Louvain Theological and Pastoral Monographs) (Leuven: Peeters, 1997); Reimund Bieringer & Didier Pollefeyt, eds, *Anti-Judaism and the Fourth Gospel* (Jewish and Christian Heritage Series) (Assen: Royal Van Gorcum, 2001); Didier Pollefeyt, *Incredible Forgiveness: Christian Ethics between Fanaticism and Reconciliation* (Leuven: Peeters, 2004); Herman Lombaerts & Didier Pollefeyt, *Hermeneutics and Religious Education* (Leuven: Peeters, 2004), Didier Pollefeyt, *Interreligious Learning* (Leuven: Peeters, 2007); Annemie Dillen & Didier Pollefeyt, *Children's Voices. Children's Perspectives in Ethics, Theology and Religious Education* (Leuven: Peeters, 2010); Marianne Moyaert & Didier Pollefeyt, *Never Revoked. Nostra Aetate as Ongoing Challenge for Jewish-Christian Dialogue* (Leuven: Peeters, 2010) and many international contributions on post-holocaust ethics, theology and Jewish-Christian relations, catechetics and religious education in a catholic setting.

Margaret Brearley read Modern Languages at Oxford University, was awarded a DAAD scholarship to Muenster University, Germany, and received her Ph.D. from Cambridge University. From 1973 she was Lecturer in German Mediaeval and Renaissance Literature at Birmingham University, taking early retirement in 1986 to found and direct the West Midlands Israel Information Centre. Brearley has held academic posts at the Centre for Judaism and Jewish-Christian Relations, Selly Oak Colleges (1987-1992) and the Institute of Jewish Affairs. London (1992-1996). She has lectured in Israel, Germany, Finland and the USA and throughout Britain, and has published many articles and book chapters, especially on anti-Judaism, genocide, the Holocaust, and Roma. Brearley founded a day centre for the mentally ill in 1979, and has been consultant to several films and documentaries, judge of the Times Preacher of the Year, honorary Advisor on the Holocaust to the Archbishops' Council (2001-4) and Honorary Research Fellow at University College, London (2006-9).

Eberhard Kolb
Bergen-Belsen
Geschichte des „Aufenthaltslagers“ 1943 – 1945

Der Name „Bergen-Belsen“erhielt 1945 die Bedeutung eines Symbols für die Untaten des nationalsozialistischen Regimes. Nahezu 50.000 Menschen aus allen europäischen Nationen fanden in den letzten vier Kriegsmonaten in Bergen-Belsen einen qualvollen Tod. Sie wurden aus anderen Konzentrationslagern hierher geschleppt, als die SS in der letzten Kriegsphase ein Lager nach dem anderen vor den vorrückenden alliierten Armeen aufgeben mußte.
Errichtet wurde Bergen-Belsen im Frühjahr 1943 als ein sogenanntes „Aufenthaltslager“ für einige trausend europäische Juden, die aus ihren Heimatländern nicht sogleich in die Vernichtungslager deportiert wurden, sondern zunächst noch einige Zeit für Austauschzwecke zur Verfügung stehen sollten. Die Geschichte dieser „Austauschjuden“– es handelte sich un einen winzigen Bruchteil der hingemordeten Juden Europas – und die Entwicklung Bergen-Belsen vom „Aufenthaltslager“ zum „Horror-Camp“ des Jahres 1945 werden in diesem Buch auf breiter Quellengrundlage dargestellt. Die Erörterung der Lagerorganisation und die Schilderung der Lager-„Gesellschaft“ führen zur Untersuchung der grundsätzlichen Züge des nationalsozialistischen Konzentrationslagersystems.
Die Darstellung wird ergänzt durch eine Dokumentation, in die außer Aktenstücken der Gestapo und des Auswärtigen Amtes auch Tagebuchaufzeichnungen einiger Lagerinsassen aufgenommen wurden, die in der Unmittelbarkeit ihrer Aussagen ein erschütterndes Bild vom Leben in diesem Lager Bergen-Belsen zeichnen.
2011, 352 S., 19,90 €, br., ISBN 978-3-643-11067-1